Presenting Technical Information

A Guide to

Presenting Technical Information

Effective Graphic Communication

Clifford Matthews

Professional Engineering Publishing Limited
London and Bury St Edmunds, UK

First Published 2000

ISBN 1 86058 249 4

A CIP catalogue record for this book is available from the British Library.

Printed and bound in Great Britain by Biddles Limited.
www.biddles.co.uk

FURTHER TITLES BY THE SAME AUTHOR

Title	**Author**	**ISBN**
A Practical Guide to Engineering Failure Investigation	Clifford Matthews	1 86058 086 6
Handbook of Mechanical Works Inspection – A Guide to Effective Practice	Clifford Matthews	1 86058 047 5
IMechE Engineers' Data Book	Clifford Matthews	1 86058 248 6

Professional Engineering Publishing (publishers to the Institution of Mechanical Engineers) offers a valuable and extensive range of engineering books, conference and seminar transactions, and prestigious learned journals, as well as a successful list of magazines. For the full range of titles published, contact:

Sales Department
Professional Engineering Publishing Limited
Northgate Avenue
Bury St Edmunds
Suffolk
IP32 6BW
UK
Tel: +44 (0) 1284 724384
Fax: +44 (0) 1284 718692
E-mail: sales@imeche.org.uk

Contents

Chapter 3: Technical Drawings

Chapter 4: Conceptual Technical Design

Chapter 5: Practical Technical Design

Chapter 8: Mechanics – dynamics

Chapter 9: Project Management Information

Chapter 10: Statistical Information

Preface (... or read this first)

> I sometimes find it a bit difficult to express technical information in a concise way. Is there any way to make this easier?

Unfortunately, the problem is wider than this – not only must technical information be concise, it must also be presented in a way that people will understand and find interesting. It is no good having an intuitive understanding of technical subjects if you can't communicate at least some of your ideas to others.

> But the variety is so wide; sometimes I have to present information using mathematics and other times using a drawing or sketch. How am I supposed to know which is best?

And sometimes you will have to use a combination of these techniques or you will find that people still get a confused message.

Sooner or later anyone working in a technical discipline will become involved in the communication and presentation of technical information. The purpose of this book is to help you in this task by showing you various ways to do it. I have tried to include sound, accepted methods of presenting information interspersed with others that need a little more imagination on your part. Technical information does not have to follow rigid presentation rules – don't be frightened to look for a new approach.

Clifford Matthews, BSc, CEng, FIMechE, MBA.

Acknowledgements

Special thanks are due once again to Stephanie Evans for her excellent work in typing the manuscript for this book.

Chapter 1

Principles

Technical information – what is it?

Technical information is information that has its roots in some sort of technique or method. It can be theoretical, practical, or a subtle mixture of the two and can be thought of as a specific form of language – the language of technology and industry. A further common factor is that technical information is related to the application of a technological skill, either in producing the information itself or using the messages that it conveys.

Where is it used?

You will see technical information used:

- in all fields of pure science;
- in all the applied sciences;
- as the bedrock of all the technical disciplines and subjects that you can think of.

Because of its wide application, the variety of types of technical information is wide and complex. Some disciplines (computer technology for example) have almost a separate technical language of their own, but the majority of technical disciplines thrive on forms of technical information that have fairly general application.

What is it for?

It is there to convey ideas between people. This is an important point;

despite the proliferation of computer-generated data of all types, the prime purpose of technical information is to convey ideas, concepts, and opinions about technical matters between people. These may be rough ideas, elusive and fleeting concepts, or finely-honed technical proofs and axioms – all come under the umbrella of technical information.

Does it have any other uses?

Yes. It is a tool of persuasion. Technical information and the way in which it is presented plays a part in convincing people of others' understanding and opinions about technical subjects. All scientific and technological activities hinge around the way that technical ideas are transferred between the participants; it is this flow of technical ideas that gives a technology or a project its direction. This means that you can think of the presentation of technical information (in all its forms remember) as perhaps *the* most common tool of science and technology. It is a tool to be used – a tool for *you*.

How to use this book

This book is intended to be useful to anyone who has to present technical information, in whatever form, to others. Its purpose is to show you how to communicate technical information in a clear and effective way, whether the information is superficial or detailed in breadth and depth. The contents are a mixture of traditional, well-tried techniques of information presentation mixed with some less well-known methods and a few new ideas. While most of the techniques are 'stand-alone' methods of presenting technical information, they are most commonly used in combination with each other. So, you can treat this book as a selection box of ready-made ideas to be chosen and adapted to fit a particular situation.

The content of this publication is not limited to a narrow set of technical disciplines. Although based mainly on technological/engineering subjects, the techniques have much wider application. It is the methods that are important, rather than the particular technical example that has been used in any individual case. Each presentation example tries to show the context in which that example is commonly used by including, where relevant:

- an example of the technique;
- its uses;
- details of variations and distortions that you may see;
- hints and tips on presentation.

The content of each presentation example varies: some lay emphasis on the precise presentation of quantitative data, while others are more concerned with the aesthetics of a presentation method and the way it can convey softer technical information about form or style. All are important parts of the subject.

I have tried to make this book one that can help you in many different areas of your technical studies and career tasks. Try to use the presentation methods widely when preparing technical reports, design studies, project submissions, or more structured documents such as technical papers and proposals. Try also to use the book to help you make the right choice of presentation methods for any application: there is rarely only a single way to present something, so aim for effectiveness by choosing the one that will show your information in the most precise way, or make the best visual impact. Finally, do not treat the presentation methods as being too rigid – they are there to be adapted or combined for your own use, as the occasion demands.

Presenting technical information – the challenge

The world of technical information is beset with the problem of complexity. The rich technical variety that exists in every technological discipline manifests itself as an ever-increasing amount of complex information that has to be presented in an easily digestible form. The task of presenting technical information is, therefore, about finding simple ways to present hard ideas. In most cases, algebraic or mathematical expressions become too complicated to be understood by anyone but purists, so it is necessary to find other ways. There are five guidelines.

SOME GUIDELINES

- Use graphical methods of communication wherever possible.
- Supplement algebraic and mathematical information with geometry to make it simpler and/or clearer.
- Use visual models to portray ideas.
- Do not be frightened to make approximations where necessary.
- Use sketches, diagrams, and drawings.

There is one common factor in these five points – all involve the use of models to present technical information effectively. The task of presenting technical information is, therefore, about constructing a representation of that information, so that its meaning can be conveyed on a computer screen or printed page.

The need for imagination

Many of the skills of effective technical presentation involve using imagination. Although traditional methods are well established, there is always room for improvement and adaptation. Trends over the past ten years have favoured the increased use of graphical and pictorial information in preference to tables of mathematical and algebraic data. Such modern presentation methods need the use of imagination to keep the development and improvement going. Imagination is also needed in the choice of presentation method to be used. It is difficult to keep technical presentations looking fresh and interesting if you use the same technique too often; you have to look for alternative ways to convey your information.

Making the choice

For any situation, when you have the task of presenting technical information you are faced with the choice of several general methods:

- tabulated (i.e. lists or tables of data);
- graphical methods;
- scientific or symbolic representation;
- technical drawings of some sort;
- pictorial representation, such as sketches and three-dimensional diagrams.

The choice between these is best helped along by learning to do a bit of critical thinking. Ask yourself a few questions about the situation such as:

- which method will help me present this technical information in the clearest way?
- is this method really suitable for this type of information, or am I just using it for convenience?
- does this method have visual power, or does it look mediocre?
- what are the positive and negative aspects of this method I am about to use?

Remember that the purpose of this type of critical thinking is to help you choose a good presentation technique, not to stifle any imagination that you are trying to bring to the process. Seen like this, the task of presenting technical information starts to resemble a process of technical problem solving: a logical choice between alternatives, coupled with a bit of imagination and flair to liven up the result.

PRESENTING TECHNICAL INFORMATION : A SUMMARY

- The purpose of presenting technical information is to convey technical ideas, facts, and opinions between people. It is also a tool of persuasion.
- There are always several different ways to present any set of technical information.
- The challenge is to find simple ways to present hard ideas. This leads to five main principles of information presentation:

  - graphical methods;
  - combining information with diagrams;
  - using 'models';
  - making approximations;
  - using sketches, diagrams, and drawings (of various types).

After all of this you have to use a little imagination and flair, and then make a decision about the best presentation method to use.

Categories of information

The way to understand better the general subject of technical information is to think of it as being divided into wide but precise categories. An understanding of the existence of these categories will also help you think critically about the purpose of different types of technical information, and how to present it in the best possible way. Figure 1.1 shows the situation. Note the three main categories: guidance only, symbolic/schematic, and prescriptive. All three are capable of belonging (at the same time) to categories of information that can be described as being inductive or deductive. Now look at the categories shown in Fig. 1.1 in turn.

Guidance-only information

Not all technical information is presented in a form that provides an exact description of something (an object, procedure, or idea). Its purpose is merely to give you guidance; to convey a general technical idea. To do this, the method of presentation often involves approximations about:

- fundamental relationships between, for example, technical procedures, designs, or physical objects;
- trends in size or movement;
- the physical shape and layout of objects or components;
- dimensions.

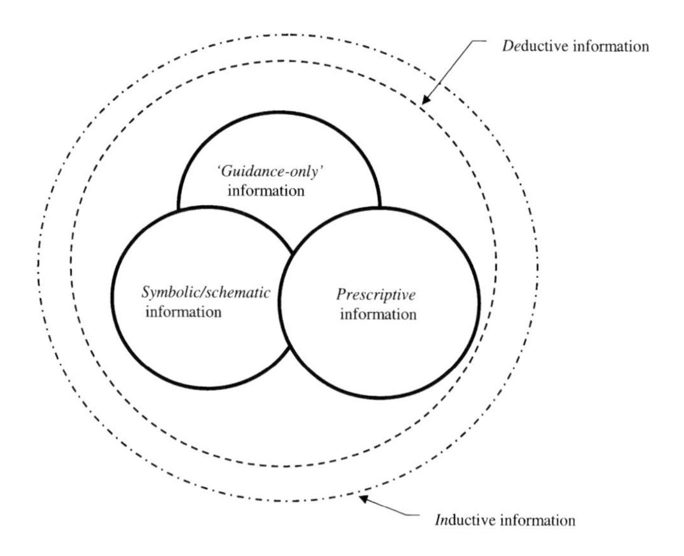

Fig. 1.1 Different 'categories' of technical information

The best way to understand this is by looking at two examples. Figure 1.2 shows a graphical representation of a simple cantilever beam supporting a suspended weight. From a quick study of the figure you should be able to infer three main points.

- The diagram is intended to show you that the beam bends under the influence of the weight; precisely how far it is bending, or the exact shape of its curvature is not shown.
- The diagram shows a 'general case', applicable to all cantilever beams; there is no attempt to show the length or cross-section of the beam, its material, or the even the size of the weight on the end.
- The beam, the weight, and the physical way that they relate to each other, are represented by a drawing which has a strong resemblance to the real visual world, i.e. it is just about the way the objects look in real life. There are obviously some approximations. The beam, for example, is represented by a thin line with no apparent depth or thickness, and there are no proper mechanical details of how the end of the beam locates into the wall, or how the weight is fixed to the free end. None of these would, however, stop you recognizing the physical arrangement of such a loaded beam if you saw one, so the drawing is a close representation of the real-life object rather then being merely a symbol.

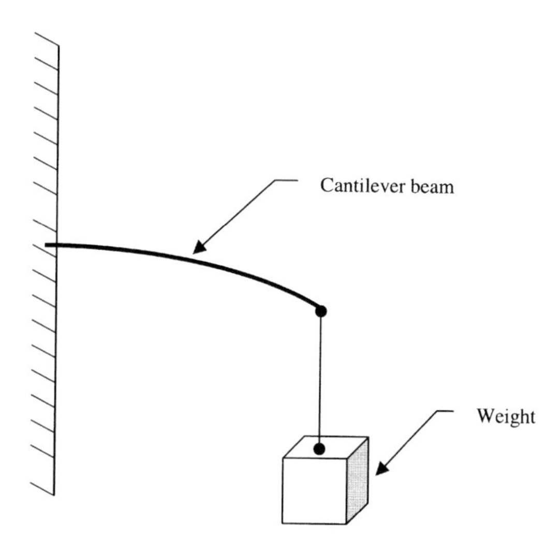

Fig. 1.2 An example of 'guidance-only' information

In summary, the message that this diagram does provide, i.e. that cantilever beams bend when a weight is applied, is only really of use as 'guidance only'. It is a non-precise but important part of the technical picture; not exactly correct, but good enough. We will see later that this type of information can be useful in many areas of technology, particularly in engineering design disciplines where technical ideas are developed in a series of steps.

Symbolic/schematic information

Symbolic and schematic types of information are so closely related that they are best thought of as a single category.

WHAT IS A SYMBOL?

A symbol is something that represents something else by association, resemblance, or convention.

WHAT IS A SCHEMATIC?

A schematic shows the scheme or arrangement of things, normally by using symbols to artificially reduce complexity.

In practice, many methods of presenting technical information are a combination of the symbolic and schematic approaches. This is valuable in just about all technical disciplines as a way of simplifying a complex system, object, or set of technical relationships down to a level that a reader can understand. In many cases, symbolic/schematic representations are the only way to portray complex technical information in a user-friendly form. Typical examples are:

- process instrumentation diagrams (PIDs) for any type of process plant;
- hydraulic, pneumatic, electrical, and similar circuit diagrams;
- applications where it is necessary to show the structure of something or how it works (such as Fig. 1.3);
- symbolic illustrations (see Fig. 1.4) which portray technical information and look attractive.

One common thread running through schematic representations is that they show directly, or infer, physical interrelationships between parts of things, often in the form of a schematic plan or design. In contrast, pure symbolic representations (as in Fig. 1.4) can be more 'stand alone', or may simply give a small piece of technical advice.

A dynamically stabilized platform

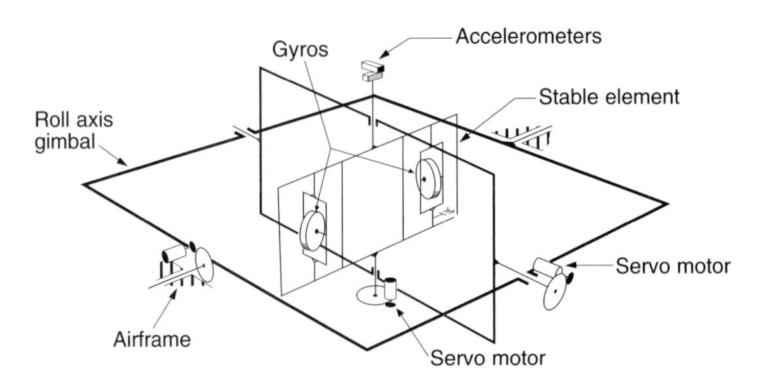

Fig. 1.3 An example of schematic information

Prescriptive information

Prescriptive information is that which sets down firm rules, or provides an exact description of something. The word itself contains the noun 'prescript', meaning a direction or decree. Not surprisingly, technical information which is prescriptive is generally complex, because it is not always possible to describe fully complex things in a simplified or shortened way. It can also have an air of rigidity about it, rooted in the fact

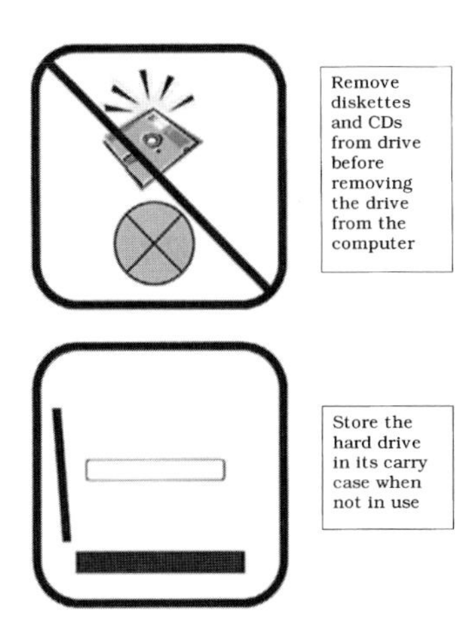

Fig. 1.4 An example of symbolic information

that it is attempting to explain the unique and detailed solution to a difficult technical situation or problem.

The power of prescriptive information lies in its ability to cause people to take action, to evaluate a system in a particular way, or to assemble a series of engineering components in the correct order. Prescriptive information is often seen (and used) 'nearer' the single solution of technical problems. This is in contrast to guidance information, which is more of an upstream technique used during earlier stages where the technical atmosphere is more diverse and conceptual.

PRESCRIPTIVE INFORMATION

Expect to see it:

- in precise mathematical routines and algorithms;
- in manufacturing procedures;
- in instruction manuals;
- anywhere where technical information contributes to step-by-step problem solving.

A further feature of prescriptive information is its accuracy. Unless technical information is accurate in number, expression (as in mathematical or algebraic notations), and in form (i.e. shapes or spatial representation) it cannot really be prescriptive, because it would leave too much freedom, thereby hindering the achievement of a unique pattern or solution. This is why prescriptive information is particularly suited to technical and engineering disciplines – it thrives on hard-edged ideas.

Deductive versus inductive information

You can think of deductive and inductive information as features of the technical background against which various presentation techniques are applied, rather than discrete presentation mechanisms in themselves. A particular set of technical information may be predominantly inductive, deductive, or (more likely) a subtle combination of the two, with at least part of the definition coming from an understanding of how that technical information was derived rather than its effectiveness in conveying technical ideas. In short, this means that you only need to consider the deductive versus inductive qualities when you come to 'fine-tune' presented information; you do not need it in the earlier stages.

INDUCTIVE INFORMATION is that which infers a future conclusion based on previous (historical) information or happenings. Examples are:

- statistical process control (SPC) in manufacturing, where the characteristics of components that are not yet manufactured are inferred by previous observation of similar already finished ones;
- most empirical laws (e.g. in fluids or mechanics) in which we draw conclusions about a large group of things from observations of one or two specific cases.

DEDUCTIVE INFORMATION has a clear link between some previous statement (called the premise) and the deduced information (or conclusion) that is presented. If the premise is true then it is deduced that the conclusion must also be true. Compare this with the inductive situation where the premise may give support to the conclusion but does not guarantee it. Common examples are:

- mathematical and algebraic expressions: i.e. $x + x = 2x$. Here x is the premise of sorts, and $2x$ is the conclusion that is obtained when x is added to another x;
- engineering drawings are primarily deductive because they describe (and so rely on) tightly controlled physical relationships between mechanical components determined before the drawing is produced.

You should now be able to see how the differences between inductive and deductive information can be built into the way that technical information is presented. Technical theories, alternatives, and concepts are suited to the use of inductive information, because it is never intended that the information is absolutely traceable to a proven premise. Think how this applies to chemistry, materials science, and the gas laws. Newton's laws of motion are also empirical (they have no proof as such) so dynamics and kinematics are disciplines in which technical information is presented in an inductive form. However, once these routines are applied for use in engineering disciplines, and are metamorphosed into engineering designs, the technical information becomes heavily deductive, as it takes its place in the search for precise solutions to engineering problems. Remember that these definitions are only relevant when presentation methods for technical information become heavily refined; they have little relevance to the simpler (often 'guidance only') forms of presentation.

Graphical methods

The term 'graphical' refers more to the way that information is presented than the nature or purpose of the information itself. Some types of technical information have a character that makes them particularly suitable to being displayed in graphical form. Graphs are best at showing relationships. These may be tight algebraic relationships linked by rigid constants and coefficients, or softer more inferred ones providing information in the form of general guidance and trends. The power of graphical methods lies in their ability to provide answers to several questions at once. A single graph can, if correctly constructed, hold information about:

- linear and non-linear relationships;
- equalities;
- inequalities;
- relationships in time and/or space;
- looser concepts such as regression, correlation and trend.

Because of the complexity that can result, graphical presentations need to be properly ordered if they are to communicate their information clearly.

The character of graphical presentation allows for a wide variety of different types of graph, but brings with it the corresponding disadvantage of an equally wide variety of distortions and misinterpretations. The effective visual impact of graphs means that it is easy to show information in a way that is capable of misinterpretation. You can also make it persuasive or misleading, if that is what you want. We will look in some depth in Chapter 2 at the presentation of graphs and, in other parts of the book, learn a little about their wide application to the job of presenting technical information.

Conventions

Conventions play a pivotal role in the presentation of technical information. They are used in both algebraic and graphical methods to bring uniformity to the way that information is presented (and interpreted) while still allowing a degree of flexibility. Do not confuse this with a set of rigid rules, which also bring uniformity, but at the expense of variety and imagination. The conventions themselves are simple; you can think of them as lowest common denominators of the presentation techniques. They are:

- scalar methods
- vector methods
- matrix methods
- dimensions

Scalar methods

Scalar methods use quantities (scalar quantities) that have a single 'real number' dimension only, normally magnitude or size. Any presentation technique that compares information on size only can, therefore, be loosely referred to as a scalar technique. Figure 1.5 gives some examples. Scalars have the following advantages:

- they are simple;
- quantities can be added, subtracted, and compared, using algebraic methods such as addition and subtraction.

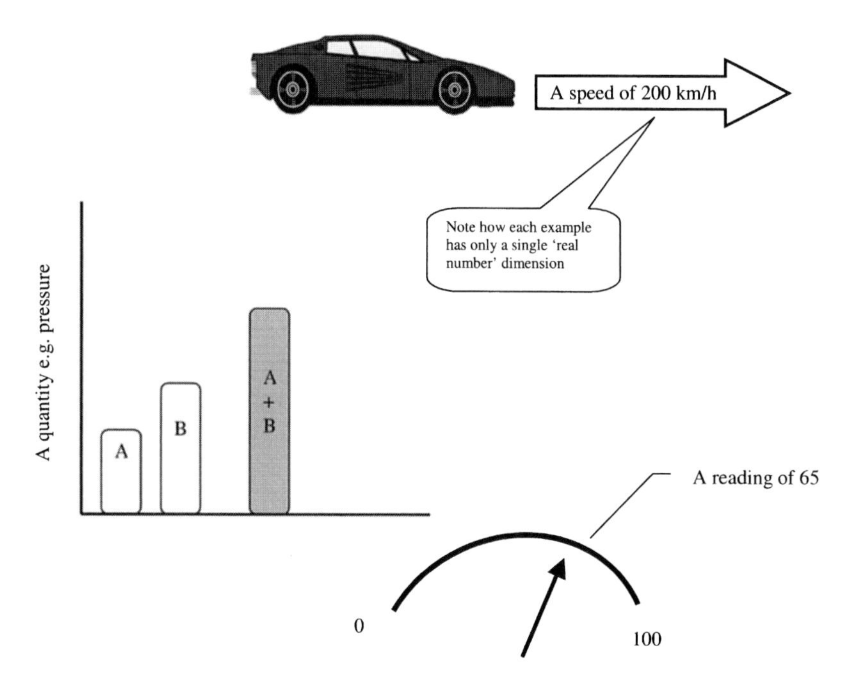

Fig. 1.5 Examples of scalar methods

Vector methods

Vector methods have more than one dimension: normally size *and* direction. The appearance of this second quantity is important, as it creates the conditions for illustrating multiple types of information about the subject that is being presented. Figure 1.6 shows some examples. Vector methods are:

- detailed (or can be);
- useful for showing complex technical situations in many different technical disciplines;
- more difficult to compare with other forms of information – you will not always be comparing 'like with like'.

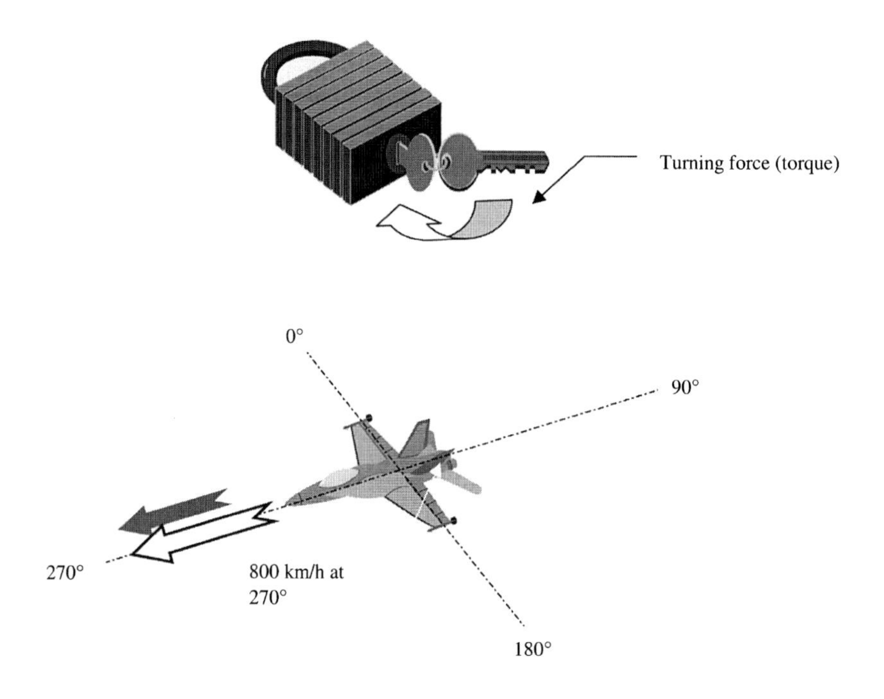

Fig. 1.6 Examples of vector methods

Matrix methods

A matrix is simply a particular type of framework in which information is contained. It is commonly used to represent a system of mathematical equations containing several unknowns. This forms part of the subject of matrix theory which, together with linear algebra, is used to present information and solve problems in disciplines such as pure mathematics, analysis of structures, thermodynamics, and fluid mechanics. Such matrices take the form of an array of 'elements' enclosed in brackets (Fig. 1.7(a)).

Matrices are also used in their more general sense to display technical information that can be contained in an arrangement of rows and columns. They can exhibit qualitative data about things that have multiple properties and are particularly useful for use as a selection tool in the design process. Figure 1.7(b) shows a typical example.

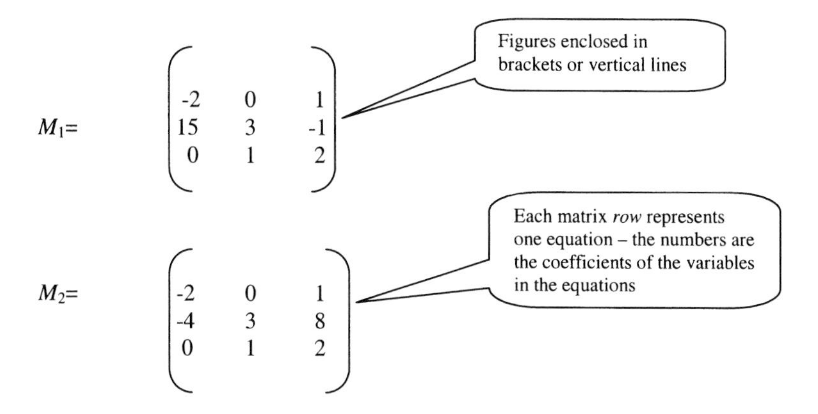

Fig. 1.7(a) A set of mathematical *quantitative* matrices

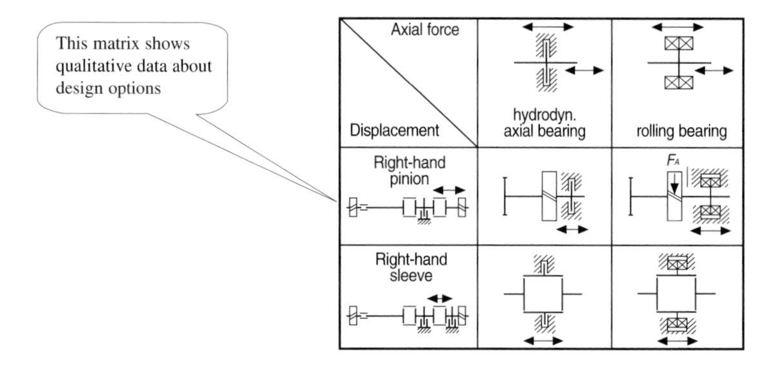

Fig. 1.7(b) A *qualitative* use of a matrix

Dimensions

It is convention that graphical and pictorial information can be presented in either one-dimensional (1-D), two-dimensional (2-D), or three-dimensional (3-D) form.

One-dimensional forms such as simple line graphs often look like they are 2-D format but in reality only convey a single 'dimension field' of information (see Fig. 1.8(a)) which could, if necessary, be conveyed by single lines. Therefore, by definition, 1-D information is capable of being conveyed by the use of simple lines of negligible thickness (Fig. 1.8(b)).

Two-dimensional information conveys information relevant to either two spatial dimensions (x and y axis, for example) or to two alternative 'dimension fields' (see Fig. 1.8(c)). Most diagrammatic and pictorial information is presented in 2-D format, so it has wide application across the technical and engineering disciplines. Two-dimensional presentations are also useful in that they can masquerade as 3-D views in applications such as wireframe drawings (see Chapter 5).

Three-dimensional presentations are used to portray pictorial views of technical objects (see Fig. 1.8(d)). Opinions vary on which type of 3-D view gives the most accurate representation. We will see later in Chapter 3 that there are several different conventions on showing 3-D views of objects, each with their own advantages and disadvantages.

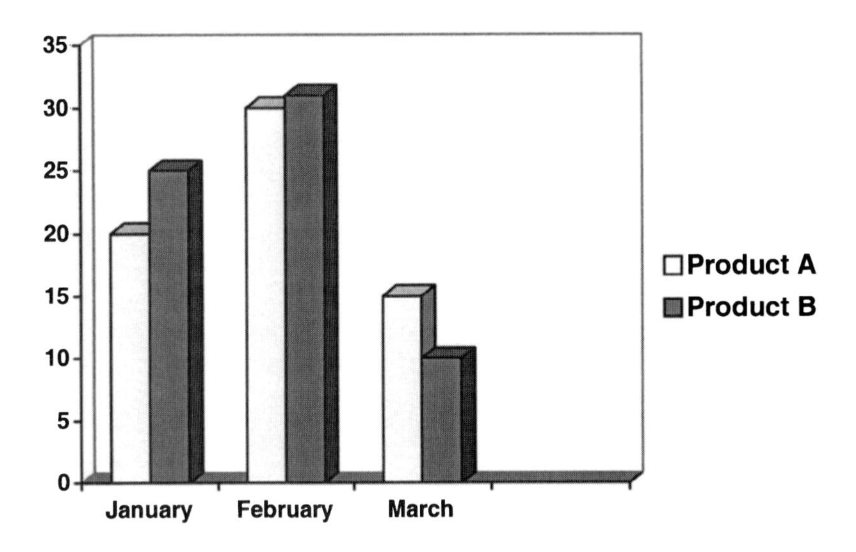

Fig. 1.8(a) This looks like 2-D or 3-D information

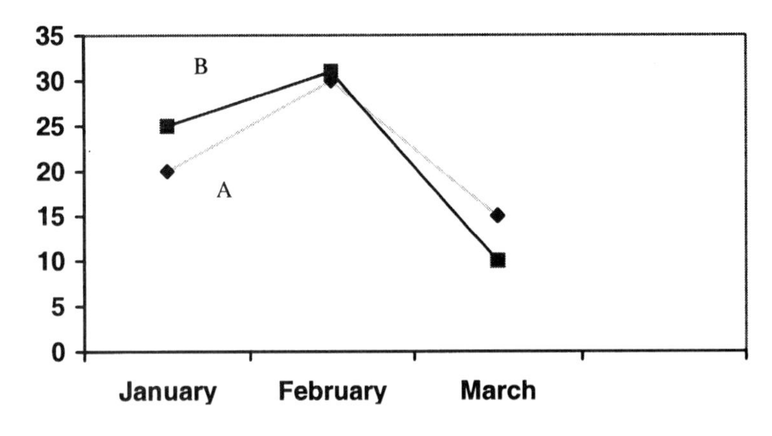

Fig. 1.8(b) The actual 1-D message of Fig. 1.8(a)

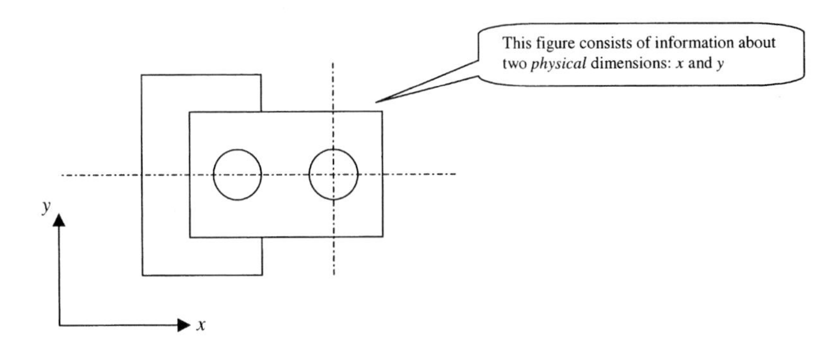

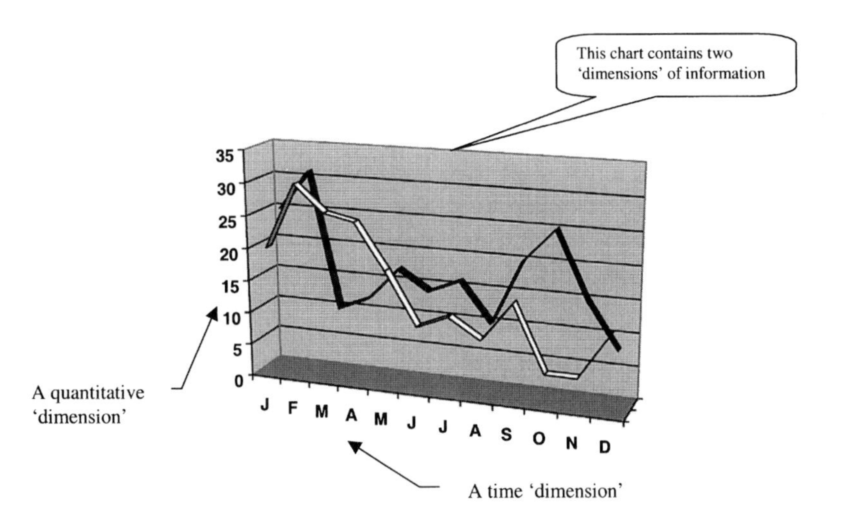

Fig. 1.8(c) Different types of 'two-dimensional' information

REMINDER: CONVENTIONS

Conventions act as the unwritten rules of technical presentation. They bring a level of uniformity to the way that information is shown. The main ones are:

- scalar versus vector presentation;
- matrix conventions;
- 1-D, 2-D, and 3-D methods.

Remember that these conventions apply to all forms of technical presentations, not just simple ones in which the conventions may be instantly apparent.

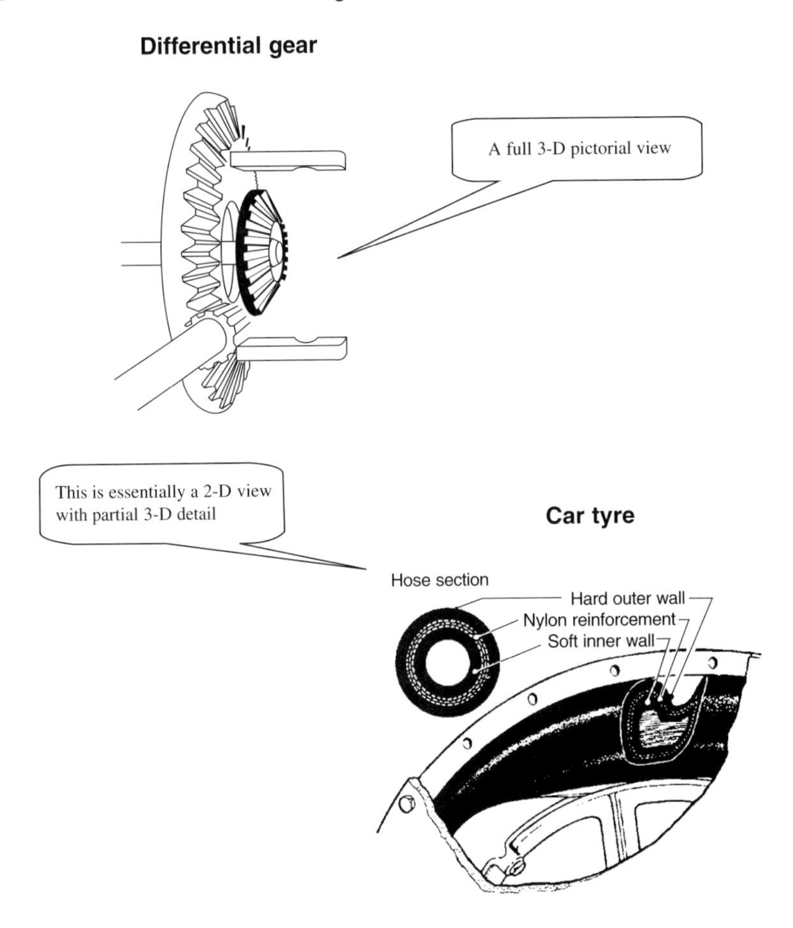

Fig. 1.8(d) Different types of 'three-dimensional' information

Co-ordinates

Co-ordinates are a method used to locate the position of points, lines, and objects in space. They are relevant to most forms of technical presentation that involve accurate graphs or drawings. Figure 1.9 shows the two main co-ordinate systems; note that they can be expressed in either 2-D or 3-D form. The cartesian system using x, y, z axes, and their positive/negative sign conventions is more commonly used for 3-D application than the 'polar' system, which is easier to depict when limited to use in two dimensions. Readings in either of these two systems of co-ordinates can be easily converted to the other. Note, however, that the fundamental differences in their sign convention means that they are rarely used together.

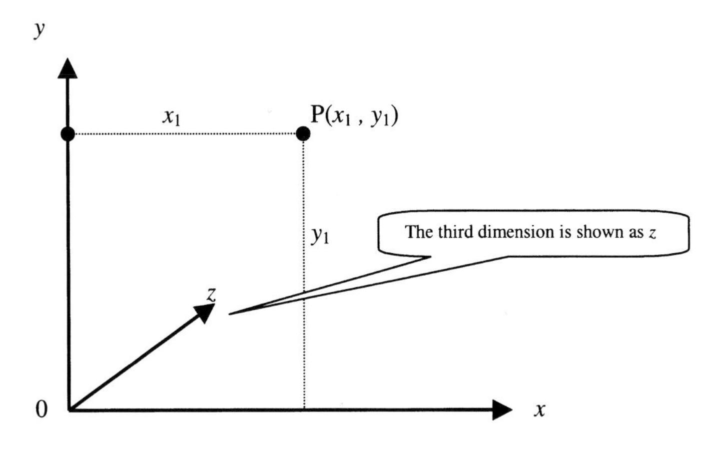

Cartesian co-ordinates

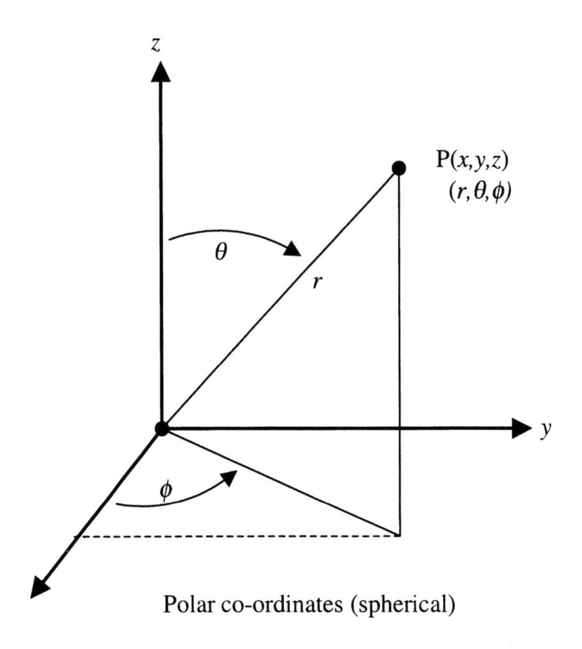

Polar co-ordinates (spherical)

Fig. 1.9 The main co-ordinate systems

CHAPTER 1: PRINCIPLES – KEY POINT SUMMARY

- Technical information has its roots in some sort of technique or method.
- It is used in all technical subjects (see Fig. 1.10).
- Good presentation of technical information involves understanding the traditional methods and then applying some imagination.
- Technical presentation is about choice. There are often several different ways to show the same thing.
- The important categories of technical information are:

  - guidance only;
  - symbolic/schematic;
  - prescriptive;
  - deductive and inductive;
  - graphical methods.

These categories are often cross-linked and combined.

- Do not forget conventions such as scalar, vector, and matrix methods, and the two main co-ordinate systems: cartesian and polar.

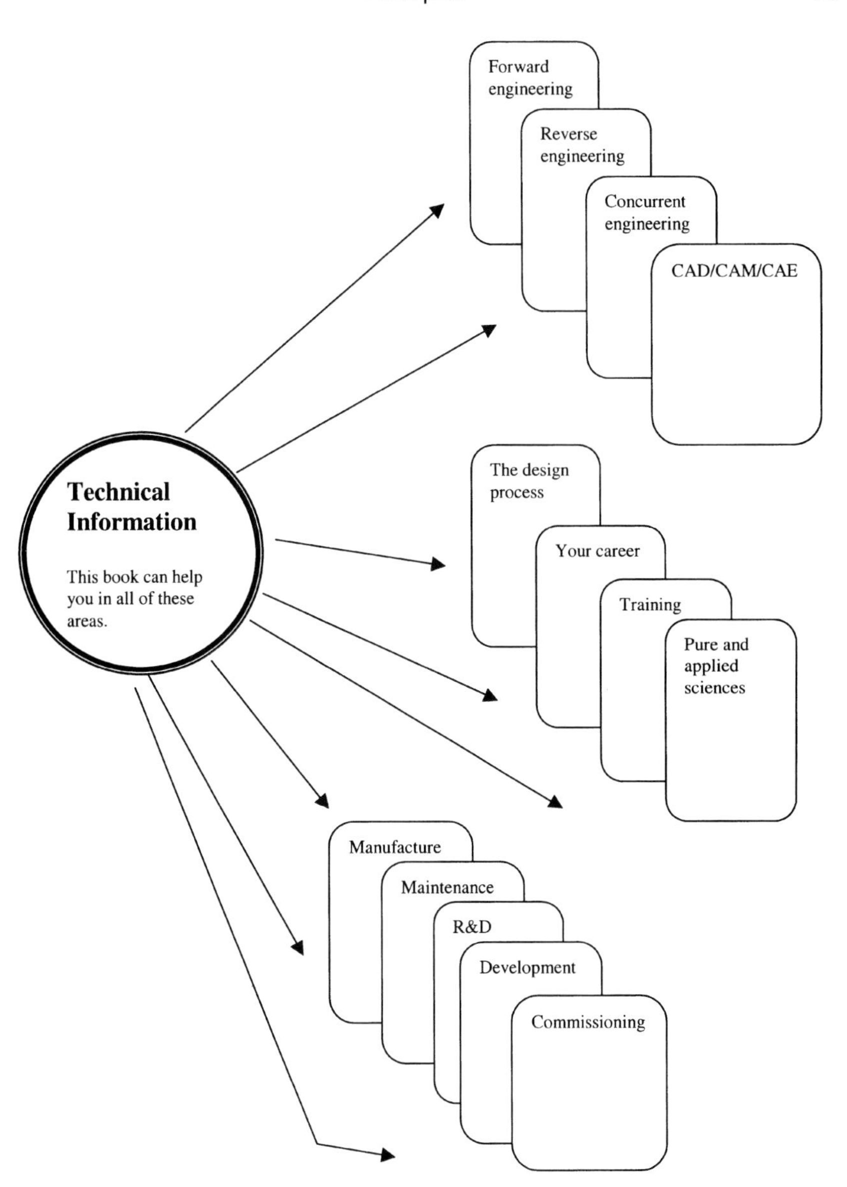

Fig. 1.10 How this book can help you

Chapter 2

Basic Data

Introduction – tables versus graphs

Just about any technical information that can be presented in tabular form can be presented as a graph, and vice versa. This does not mean that tables and graphs are interchangeable, merely that they use the same source data in their formation.

Tables are useful when:

- data values are widely spread;
- it is only necessary to compare like with like (i.e. there is a single type and character of data);
- the amount of data is small – less than about 25 entries.

These points demonstrate the common limitations of presenting technical information in a table, the main one being that it is difficult to assimilate large amounts of data at first glance. This makes features such as trends and overall patterns inherent in the data difficult to spot without detailed study. Tables also suffer from the disadvantage that they can only hold a single 'field' of data so multiple tables have to be used in order to portray a complex technical picture. Figure 2.1 illustrates this.

This table is comparing data about essentially different types of product, so *conclusions* are difficult

Type of pump	Q (m^3/s)	H (m)	Speed (rpm)	Weight (kg)	Models in range
Centrifugal	1–5	<30	200-5000	100–700	5
Rotary	4–30	<30	300-5000	150–800	6
Screw	1–6	<70	200-6000	25–300	3
Multi-stage	1–20	<80	<6000	50–400	Bespoke
Single-stage	1–20	<23	<6000	30–400	25
Reciprocating	1–7	<100	<300	100–800	2

Note how it is very difficult to see the overall 'picture' from these data

Fig. 2.1 The problem with tables

Graphs are better because they make technical information easy to grasp. Data points and quantities that would be difficult to assimilate from tables can be clearly displayed in graphical form.

GRAPHS ARE A PARTICULARLY POWERFUL WAY TO PRESENT:

- An instant 'snap shot' view of a situation involving discrete data entries such as spatial dimensions or financial figures;
- **Relationships between numbers**;
- **Equalities and inequalities,** particularly where these have robust mathematical definition;
- **Trends,** i.e. whether numbers are increasing, decreasing, or following some kind of pattern;
- Correlation between one set of information and others and projections, i.e. events or situations that have not happened yet. Typical examples would be graphs that show asymptotes (curves that approach but never reach other lines or axes). These are in common use in mathematics, economics, electronics, aeronautics, and other disciplines involving calculus;
- **Time-series data,** i.e. how things change as time progresses. This type of data is often bivariate (containing two separate sets of values) which makes it difficult to assimilate from a table.

These core differences between the characteristics of tabular and graphical methods of presenting information extend across the technical disciplines. While tabular information still proves its usefulness in areas where high accuracy is required (graphs are often not that accurate), graphical techniques are ever increasing in popularity. This is assisted by the rise in user-friendly computer packages and a more general trend towards visual methods rather than 'dry' lists and tables.

Using graphical methods

The secret of graphical methods is in knowing how to use them. You have to choose a method that can show a set of technical information in the best way, and then tailor that method to your specific requirements. There are two main choices to think about:

- the type of data (whether it is discrete or continuous);
- the type of judgement that will be made from the graph, i.e. whether its purpose is to:
  - summarize data;
  - compare data;
  - describe the interrelationships between data.

The result is that there is a large variety of individual graphical presentations, any of which can be combined with supporting tabular data, 2-D drawings, and pictorial diagrams to provide an almost infinite number of possible presentation methods. There is no need to be wary of this. Think of it as providing you with a rich variety of ways in which to express yourself, providing a vehicle for your technical imagination rather than forcing you into a search for a single predictable solution. Single presentation solutions do not exist – there are always several ways to present the same thing.

Scale charts

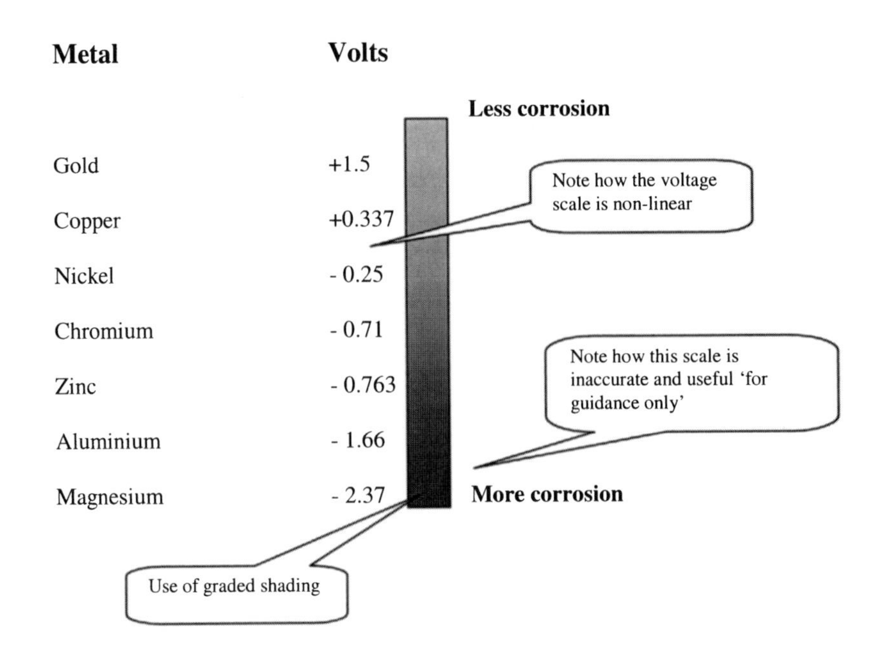

Fig. 2.2 An example of a scale chart

Uses

Scale charts are used to display a single field of technical data, particularly when the data are arranged on a non-discrete or 'sliding' scale. Each data entry has a corresponding numerical value.

Particularly suitable for:

- single-dimension physical quantities such as lengths, wavelengths, chemical concentrations, etc.

Data and units

The data are presented in tabular form so there are no axes, as such. Any units can be used as long as they are numerical (or at least quantitative).

Variations and distortions

- Watch out for non-linear scales, which can make the entries appear more closely related than they really are.
- Some scale charts have more than one vertical scale (as in Fig. 2.2). One of these scales may not be accurate and so is suitable for 'guidance only'.

SCALE CHARTS – TECHNICAL TIP

Scale charts often use graded shading (as shown) to emphasize that the data exist on a sliding scale rather than as discrete categories.

Vertical box graphs

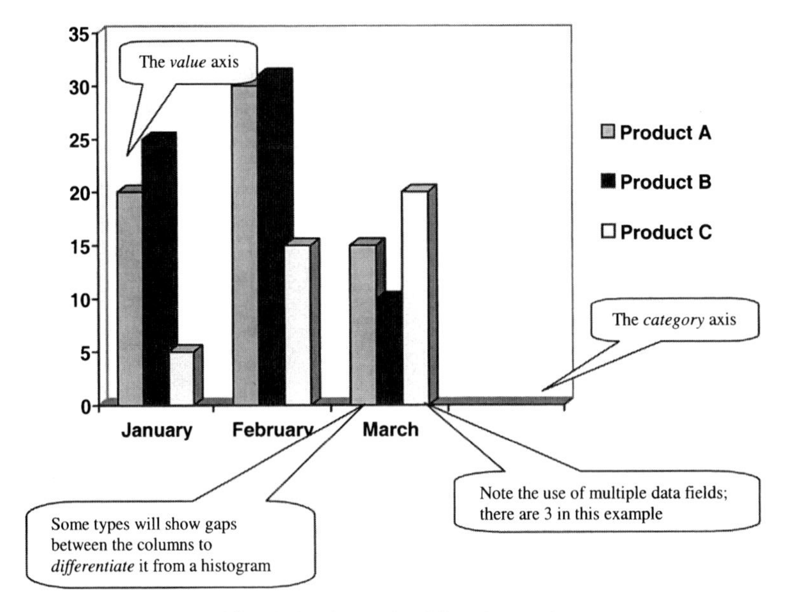

Fig. 2.3 A vertical 'box' graph

Uses

Often referred to (incorrectly) as histograms, these are used anywhere where it is necessary to present a visual impression of differences between discrete data groups. They allow you to make a measured comparison by scaling directly off the column lengths.

Particularly suitable for:

- data related to production of components and products;
- time-series data, i.e. displaying a collection of data values measured in discrete intervals of time (as in Fig. 2.3).

There are many different types of vertical bar graphs: they can be thought of as a family of presentation techniques rather than a single type.

Other uses:

- presenting multi-series data (product sales, etc.) on a single graph; in practice, most vertical bar graphs present multi-series technical information;
- presenting bivariate data – these are data where each entry represents a pair of data values rather than only one. Figure 2.3 shows bivariate data because each entry shows the number of products made and the month in which they were made.

Data and units

- Quantities such as numbers, values, costs, and dimensions are generally shown on the vertical ('value') axis.
- Time is nearly always portrayed on the horizontal ('category') axis. Time units are discrete (seconds, hours, months, etc.). The horizontal axis must contain discrete categories (or the graph becomes a histogram).
- Axes may have gridlines that make it easier to take quantitative readings from the graph.

Variations and distortions

The most common distortions in the basic vertical bar graph centre around the way in which the axes are scaled (usually the vertical one). Watch out for:

- shrinking of the vertical axis so it represents very small changes and, therefore, gives a misleading impression of the differences between the heights of the columns;
- axes that do not start from zero. This again, gives a misleading visual impression of the data.

VERTICAL BOX GRAPHS – TECHNICAL TIPS

- Keep to a maximum of three separate data series (sets of columns) on a black and white graphical presentation. If you need more, use colour, or even two separate graphs – to avoid confusion.
- Make the vertical columns the same width; this allows a fair comparison using column height alone.
- Leave gaps between the columns; this emphasizes that the horizontal axis comprises discrete data categories rather than a continuous numerical series (histogram).

Useful reference

http://www.data-surge.com/datasurgecorpo/grapher.htm

Vertical 'histogram' graphs

Uses

This is the first step in adapting the standard vertical 'box' graph for statistical application. The peak values of the columns are used as data points for the formation of a distribution curve that can be subject to statistical analysis.

Particularly suited for:

- manufacturing control techniques such as statistical process control (SPC);
- data series which contain a large number of samples, e.g. population data, economic indicators, financial portfolios, failure, and risk analysis.

Data and units

Quantitative readings are retained on the vertical ('value') axis. The horizontal ('category') axis is always numerical and represents adjacent intervals or a continuous number line. This allows the drawing of the distribution curve that is commonly seen in production or quality assurance statistics (see Fig. 2.4(a)).

Variations and distortions

- Insufficient data series (columns) to justify the drawing of a smooth distribution curve. This gives a misleading impression about the shape and smoothness of the curve that can accurately be defined from the column data given.
- Using a non-linear or non-continuous horizontal axis scale. This can misrepresent the shape of the resulting distribution curve, perhaps making it resemble one of the common statistical distributions (when it actually does not).
- Collapsing several class intervals together. This falsely accentuates the importance of the wider interval (see Fig. 2.4(b)).

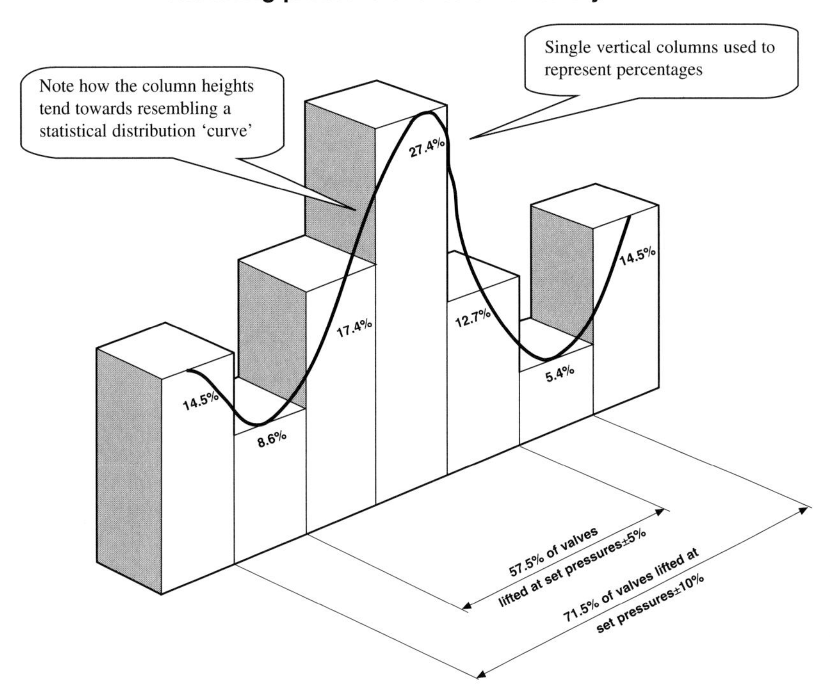

Fig. 2.4(a) A vertical 'histogram'

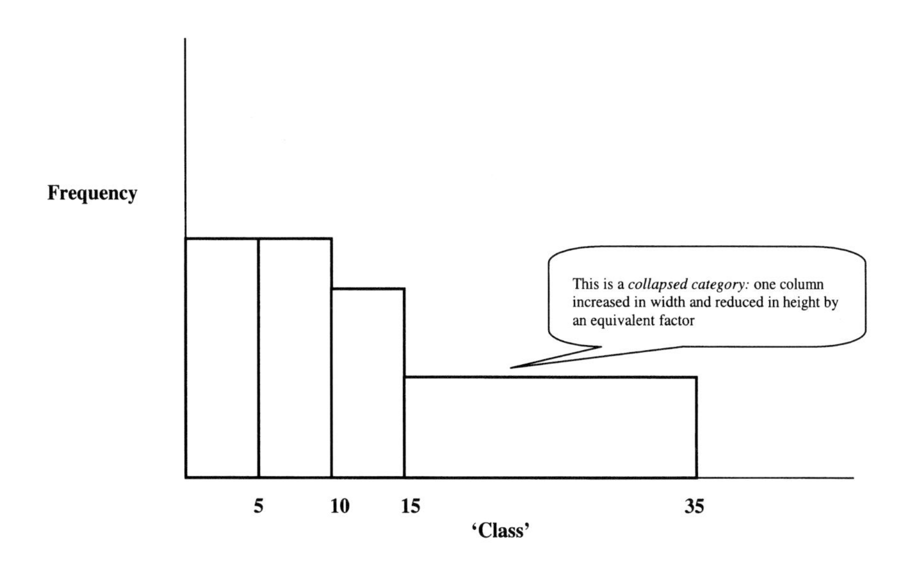

Fig. 2.4(b) An example of a 'collapsed class' histogram

VERTICAL 'HISTOGRAM' GRAPHS – TECHNICAL TIPS

- Use at least six columns to draw a distribution curve; any less and it will not be accurate (and may be misleading).
- Do not show distribution curves as dotted or chain lines; they are conventionally shown as solid lines.
- Remember that proper distribution curves need a lot of points before they are accurate enough for statistically significant conclusions to be drawn from them. It is best to present the type of distribution shown in Figs 2.4(a) and (b) as 'for guidance only'.

Horizontal bar graphs

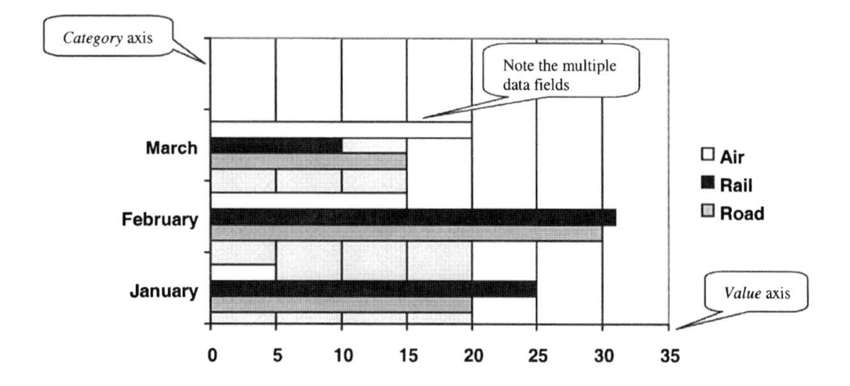

Fig. 2.5(a) A horizontal bar graph

Uses

These display the same type of data as vertical bar graphs, except that they are drawn as horizontal bars rather than vertical columns (see Fig. 2.5(a)).

Particularly suitable when:

- there are a lot of data categories, which makes it difficult to display all the category names at the bottom of vertical columns;
- the data categories have long names.

Data and units

The same guidelines apply as for vertical bar graphs, i.e. the category axis (now the vertical one) shows the discrete categories, while the horizontal axis shows numerical values. This type of presentation is often used to portray physical distances because it is conceptually easy to recognize distance as being represented 'along' the horizontal axis from left to right.

Variations and distortions

The most common variation is to show the horizontal bars split into their components; this is sometimes loosely known as a 'component horizontal bar chart' (see Fig. 2.5(b)). The advantages are:

- it allows easy comparisons between related parts of the assembled data;
- the vertical axis allows procedural steps to be shown vertically (top-to-bottom); this helps them to be recognized as chronological stages (as in the Fig. 2.5(b) example).

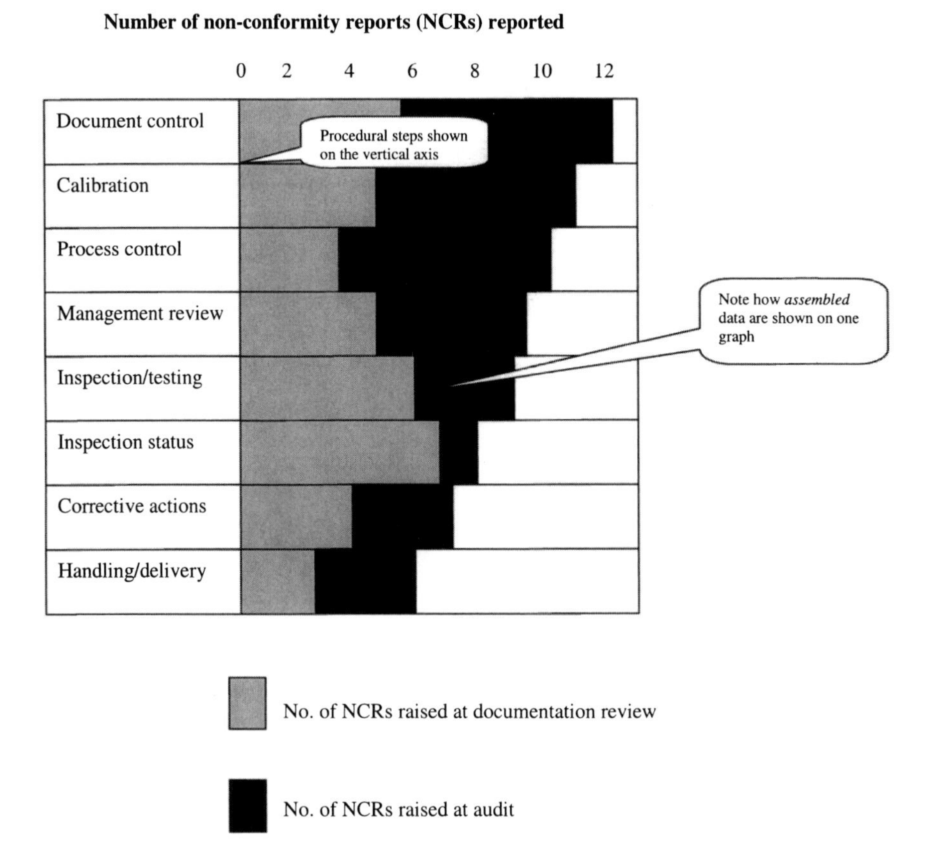

Fig. 2.5(b) A 'component' horizontal bar chart

Useful reference

http://www.data-surge.com/datasurgecorpo/grapher.htm

Line graphs – linear scales

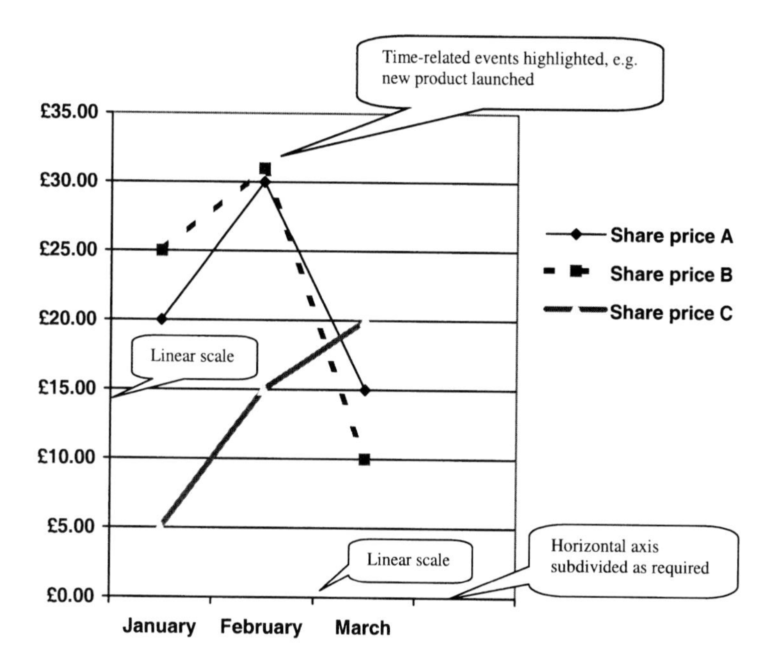

Fig. 2.6 A simple line graph with linear scales

Uses

Line graphs with linear scales are used for a wide variety of applications. In their simplest form, as in Fig. 2.6, they are used to display information about discrete data categories, with the points being joined by a straight or curved line to indicate a general trend. The alternative is to display information about continuous data: this is done by a histogram curve which is a particular form of line graph.

Particularly suitable for:

- multiple data fields (several lines on one graph);
- showing absolute values or trends (time-series data and events);
- showing design and performance data for engineering components.

Data and units

Numerical values can be shown on either axis. Numbers can be positive or negative. The horizontal axis is nearly always subdivided.

Variations and distortions

- Making line graphs by joining the tips of the columns of vertical bar charts. This is the most common distortion. The effect is to show discrete data masquerading as continuous data, which means that the resulting curve can be meaningless.
- Forming a line graph by joining the tips of columns that have been placed in a misleading order. This is a more subtle distortion, but one which can also make the shape of the line graph meaningless.

LINE GRAPHS (LINEAR) – TECHNICAL TIPS

- Do not try to mislead by mixing up discrete and continuous data – it is too easy to detect.
- Use thin plain lines for lines and curves – thick ones and fancy designs do not look professional.

Useful reference

http://ftp.sas.com/techsup/download/technote/ts304.html

Line graphs – logarithmic scales

Uses

These are a specific form of line chart which are adapted to use logarithmic scales. They are common in scientific and technical disciplines and, surprisingly, other quantitative subjects such as finance and economics. Their advantage is that they artificially shrink the spread of a set of data and so make it easy to comprehend.

Particularly suitable for:

- graphs of mathematical expressions involving logarithms, e.g. $y = \log x$ or $\log a = C(\log b)$;
- selection and sizing charts for engineering components (see Fig. 2.7).

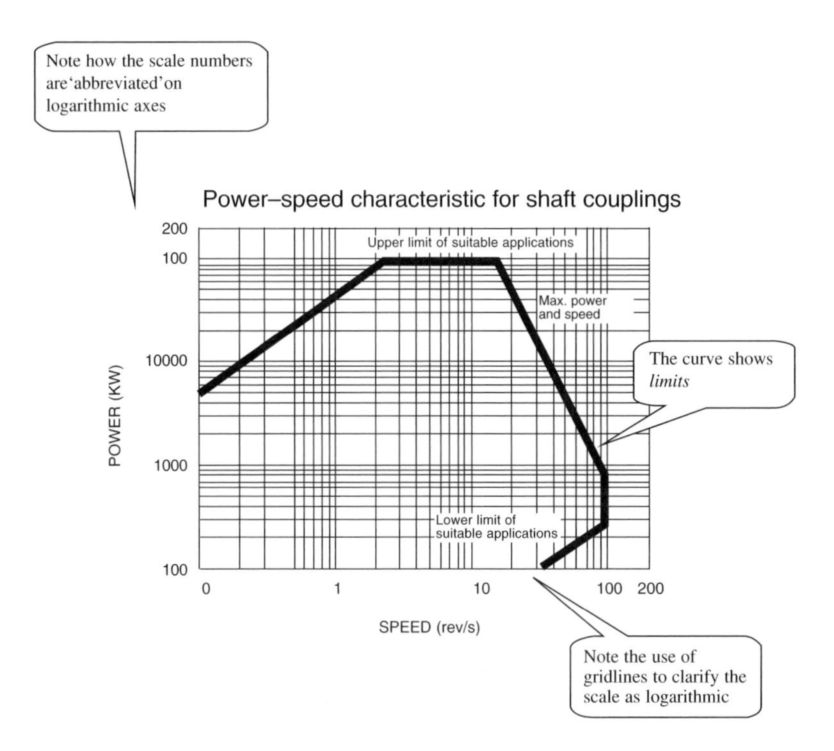

Fig. 2.7 A typical logarithmic scale graph

Data and axes

Both horizontal and vertical axes must contain numerical values. Either or both axes may be set to a logarithmic scale to any base; most tend to be $\log_{10}$ or $\log_e$ (ln).

Variations and distortions

- Using logarithmic scales when the data do not warrant it. This results in a misleading shape to the characteristic.
- Not making clear which scales are logarithmic. You may see this combined with insufficient numbers on the scales, making it difficult to see at first glance whether a scale is logarithmic or not.

LINE GRAPHS (LOGARITHMIC) – TECHNICAL TIP

- Always use gridlines on the logarithmic axis. This makes it clear that a logarithmic scale is in use. Leave linear scales free of gridlines to provide a contrast.

Useful reference

http://www.klg.com/xrt/3d/

Line charts – probabilistic

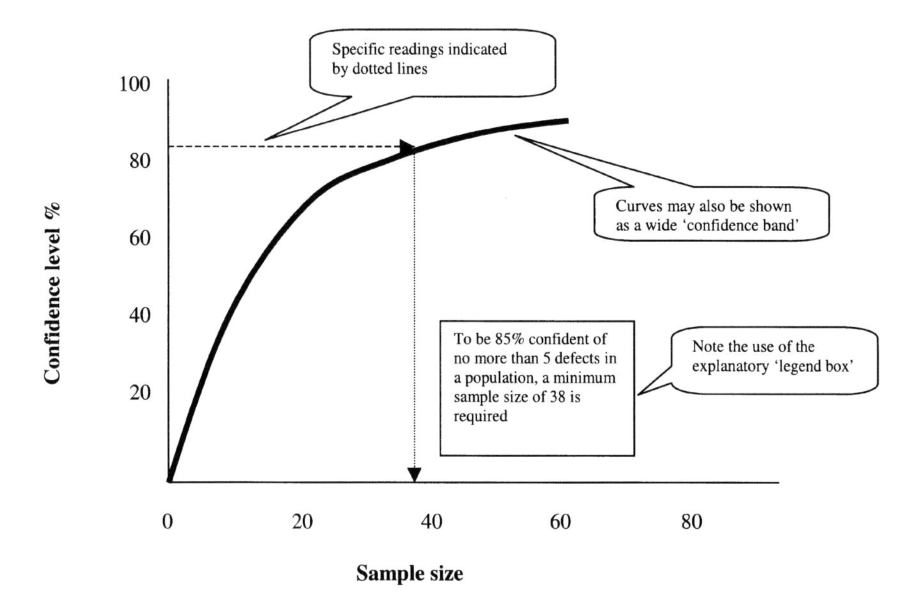

Fig. 2.8 An example of a 'probabilistic' line chart

Uses

The main application of probabilistic line charts is to show statistical information relating to confidence levels.

Particularly suitable for:

- displaying the implications of various sample sizes in research work;
- statistical process control (SPC) of mass-produced manufactured components.

Data and axes

Sample size is traditionally shown on the horizontal axis with confidence level (in percent) on the vertical axis. Two or more confidence curves can be shown on the same graph; they are often solid dotted/chain dotted to differentiate between them. The curves are sometimes replaced with shaded or hatched bands, indicating the idea of a confidence 'zone' around the readings.

Variations and distortions

- Not starting the vertical scale (confidence level %) from zero. This gives a misleading shape to the curve suggesting, on first glance, that the confidence level is different to what it actually is.
- Using non-linear axes (either one). This makes for a very misleading impression of the shape of the confidence level curve.

LINE CHARTS (PROBABILISTIC) – TECHNICAL TIPS

- Do not use gridlines on confidence level curves; it is the *shape* of the curve that conveys the main message.
- It is wise to use a legend box (as in Fig. 2.8) explaining what the curve really means. This will help prevent misinterpretation.

Line charts with 'envelopes'

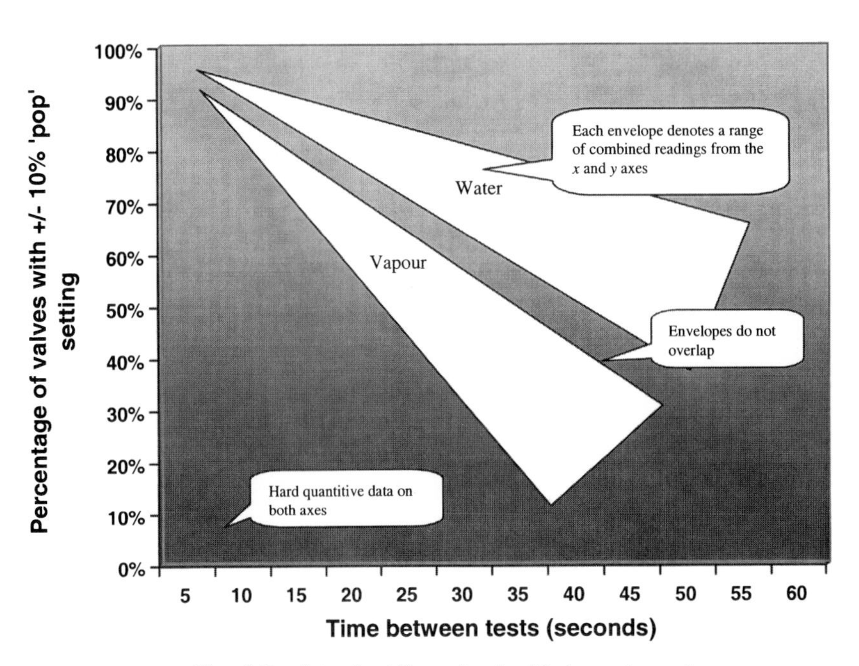

Fig. 2.9 A typical line chart with 'envelopes'

Uses

Use these when you want to show one or two *ranges* of data readings on a normal two-axis line chart. Each envelope denotes a range of data readings that have some common property.

Particularly suitable for:

- showing technical differences between engineering products or ranges of products.

Other uses:

- process engineering; showing ranges of reaction variables;
- mechanical engineering; general selection charts;
- production engineering; showing the volume or 'quality ranges' of manufactured products.

Data and units

Because of the existence of envelope regions, both axes normally represent continuous (numerical) rather than discrete (category) data. The data points are not shown, so it is *inferred* that enough lie on the lines surrounding each envelope to define its shape.

Variations and distortions

Watch out for envelopes that are surrounded by dotted rather than solid lines. This can indicate data that are extrapolated or estimated rather than being based on actual empirical results. Expect to see this in:

- charts showing the properties of proprietary materials;
- claimed performance characteristics of fluid/heat engine-based machines.

Some charts show performance envelopes that overlap each other. These are a fertile source of misleading information.

LINE CHARTS WITH 'ENVELOPES' – TECHNICAL TIPS

- Omit gridlines, it makes the chart neater.
- Do not let the envelopes cover more than about 40% of the chart area (for clarity).
- It looks better if you can arrange the envelopes to 'originate' in the same area of the chart (as in Fig. 2.9).

Line charts with 'field ranges'

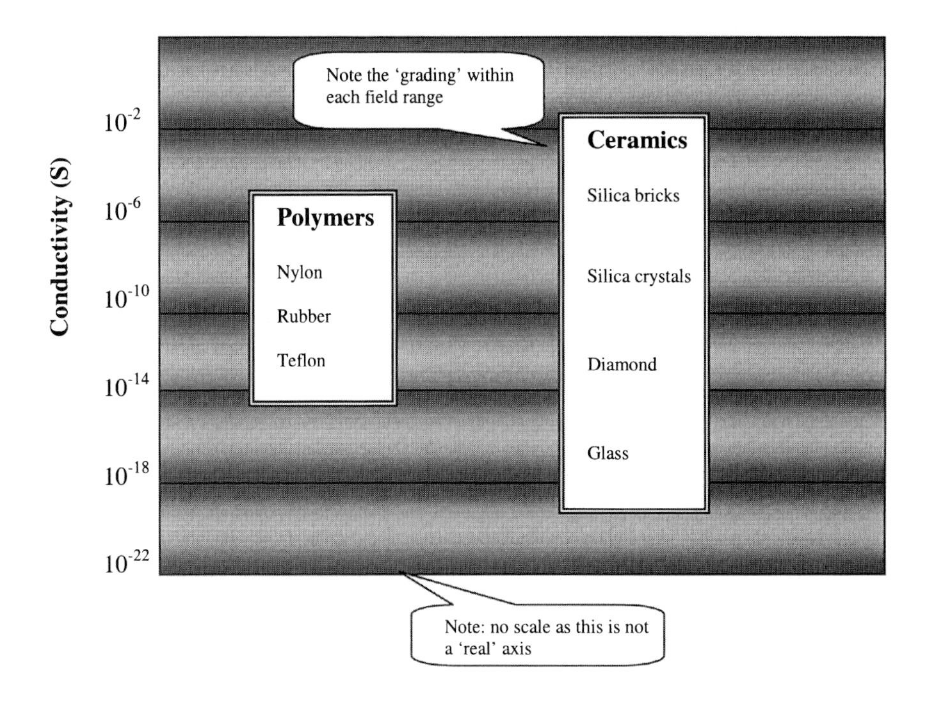

Fig. 2.10 A typical line chart with 'field ranges'

Uses

These are a derivation of line charts, used when the objective is to show different 'ranges' of information. They are often used when ranges of information overlap each other, possibly making the situation difficult to depict in other ways.

Particularly suitable for:

- closely spaced information ranges in electronics, chemistry, and materials science;
- size or performance data for ranges of manufactured components or equipment.

Data and units

The vertical axis is formatted to a linear or logarithmic scale of units, while the horizontal axis normally has no units at all. The horizontal axis is used simply for physical separation of the data ranges. The data field ranges are discrete and hard-edged; solid lines are used, inferring that the (vertical axis) data can be well defined. Note that this applies whether the data are discrete or continuous.

Variations and distortions

- Use of a logarithmic or 'broken' vertical axis to provide misleading first impressions.
- Representing the ranges by 'columns' of different widths, making some ranges look more important that others. Remember that the horizontal axis is generally a virtual one, so the widths of the field columns have no real significance.

LINE CHARTS WITH 'FIELD RANGES' – TECHNICAL TIP

- Include additional subdivisions within each field range (as in Fig. 2.10) to provide extra technical 'grading' information.

Line charts – quantitative assessments

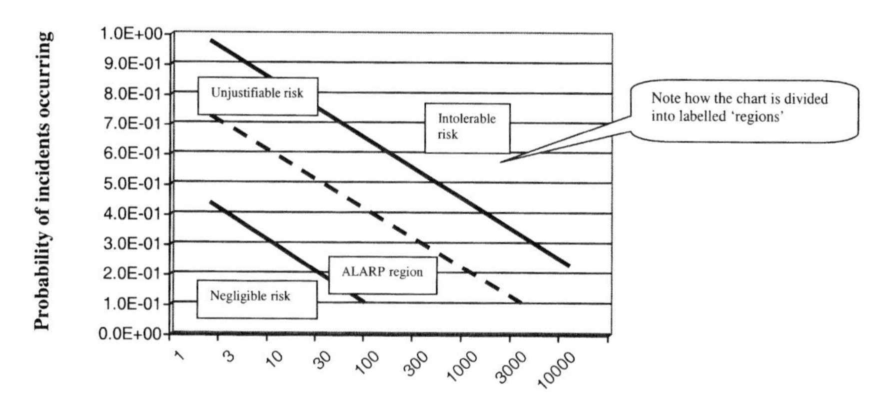

Fig. 2.11 A typical 'quantitative' line chart

Uses

These have a specific purpose, i.e. to display heavily qualitative data in a form that makes them look as quantitative as is possible. A further objective is often to divide up large amounts of empirical data by displaying them in amalgamated form.

Particularly suitable for:

- showing risks and probabilities of technical events, concerning large data populations;
- societal and demographic analyses and cause/effect consequences.

Data and units

Both axes are normally linear; it is rare to see logarithmic scales in this type of technical presentation. The chart is divided up into loose open-ended 'regions' by separate sets of lines. The regions frequently contain separate legends to explain their significance.

Variations and distortions

The most common distortion is the use of different types of lines (i.e. solid, dotted, etc.) to designate the 'regions' of the chart. This infers that there is more than one type or technical status of region or that the regions have different significance (but it does not tell you how or why).

LINE CHARTS (QUANTITATIVE ASSESSMENTS) – TECHNICAL TIP

- It is convention to use the vertical axis for the smallest units or most heavily subdivided scale (as in Fig. 2.11).

Multi-axis line charts

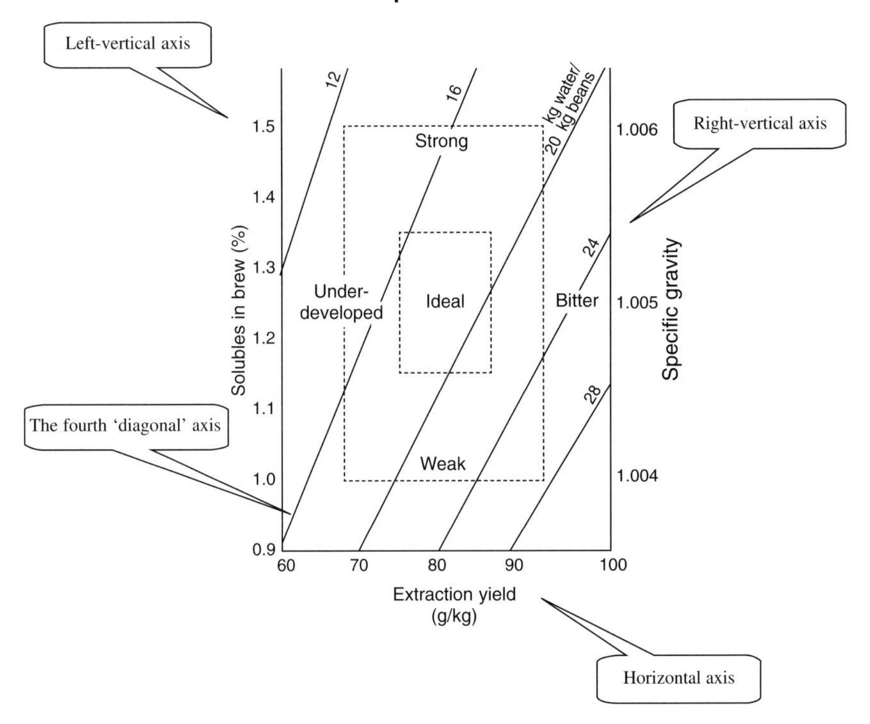

Fig. 2.12 A typical 'multi-axis' line chart

Uses

Multi-axis charts are used when you want to display more than two types of information in a single two-dimensional presentation. This is necessary in many technical disciplines, because of the sheer complexity of the subject matter.

Particularly suitable for:

- presentations in materials science, food science, and medical disciplines;
- data trying to show the quality or performance of something against its technical design data.

Data and units

Note the three scaled axes in Fig. 2.12. These are the left vertical, right vertical, and horizontal – all of which have different scales. The series of diagonal lines simulates a fourth axis, complemented by the dotted concentric boxes. These boxes effectively form the 'output' of the chart, providing the conclusions. Because of their complexity, multi-axis charts have only very broad subdivision on their three 'primary' axes and do not pretend to be highly accurate.

Variations and distortions

These charts are difficult to design from scratch because it is easy to choose the wrong scales. Expect, therefore, some level of visual distortion and inaccuracy. You have to read them carefully.

MULTI-AXIS LINE CHARTS – TECHNICAL TIP

- When designing a chart like this, work around the diagonal 'fourth' axes. Design it so these are straight, and then choose the scales of the other axes to suit. This will make the chart easier to read.

Useful reference

http://www.data-surge.com/datasurgecorpo/grapher.htm

Life cycle line charts

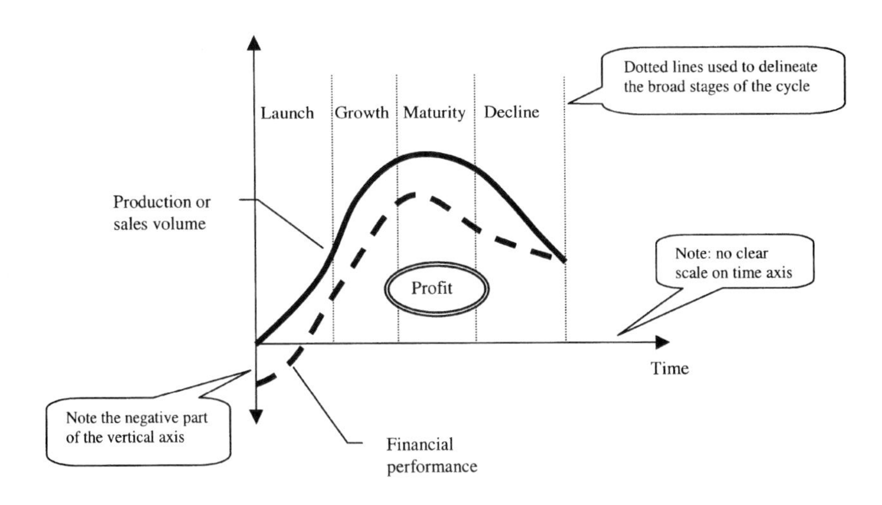

Fig. 2.13 A typical 'life cycle' line chart

Uses

This is a simple adaptation of the line chart format, used to convey the idea of a life cycle, i.e. the variation of some property of a project, product, etc. through time.

Particularly suitable for:

- product development and marketing;
- project management-related disciplines;
- reliability and failure analyses of engineering components.

Data and units

Figure 2.13 shows a typical presentation in which the horizontal axis broadly represents time scale, and the vertical axis some type of volume-related units (revenue, numbers produced or sold, etc.). The axes' units are usually kept either very basic or missed out entirely, because the main purpose of the chart is to show *trend information* about a life cycle rather than quantify it exactly. Life cycle data (unless historical) are notoriously inaccurate, so heavily subdivided axes are unnecessary.

Variations and distortions

- Watch out for vertical 'zones' of the life cycle curve being of different widths. This infers that the horizontal 'time' axes may not be linear, and there may be no graduated scale to check it against.
- Be wary of areas of the chart containing a legend. It can be difficult to know exactly which part of the total chart area the legend is meant to be relevant to. The 'profit' legend in Fig. 2.13 is a good example. Which parts of the area under the curve do you think it applies to?

LIFE CYCLE LINE CHARTS – TECHNICAL TIPS

- Try to size the vertical axes so that the life cycle curve extends well into the top half of the chart area; flat curves look inaccurate, even if they are not.
- Do not use shading in life cycle charts; it is the curve that has the significance, not the areas above and below it.

Illustrated line graphs

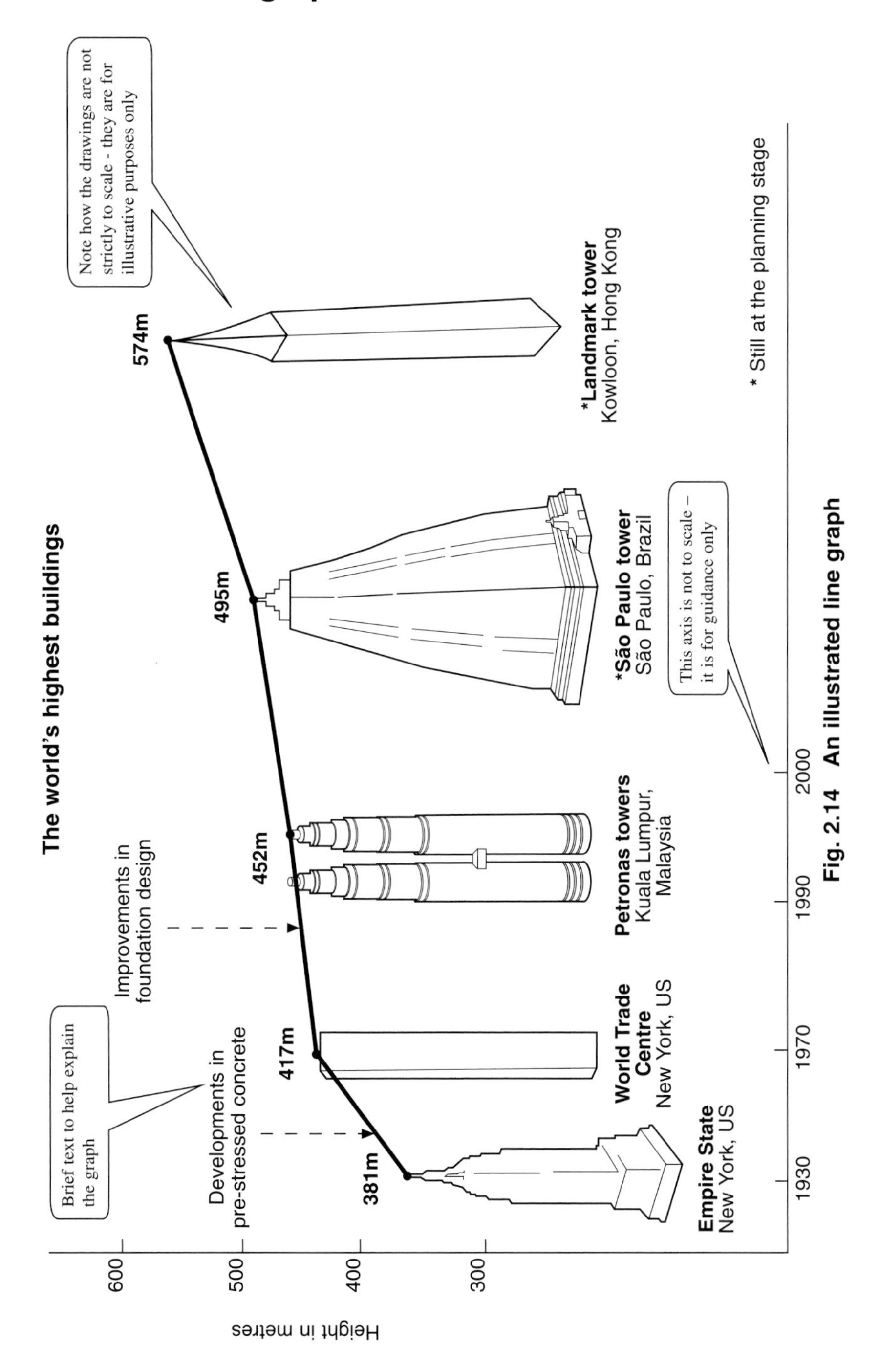

Fig. 2.14 An illustrated line graph

Uses

These are pictorial or semi-pictorial versions of line graphs, normally restricted to the two-axis type. Their main objective is to give visual interest to otherwise rather dry technical subjects (see Fig. 2.14).

Particularly useful for:

- 'popular' science and technical subjects intended for a less specialized readership;
- applications which encourage the use of colours, silhouettes, or other methods of obtaining a quick visual impact on the reader.

Data and units

The key point is that the illustrations used must not compromise the accuracy of the chart, even though this is primarily not a method of presenting technical information accurately. The scales may be shown in an approximate or even pictorial form; this is normally limited to one axis, to keep things simple.

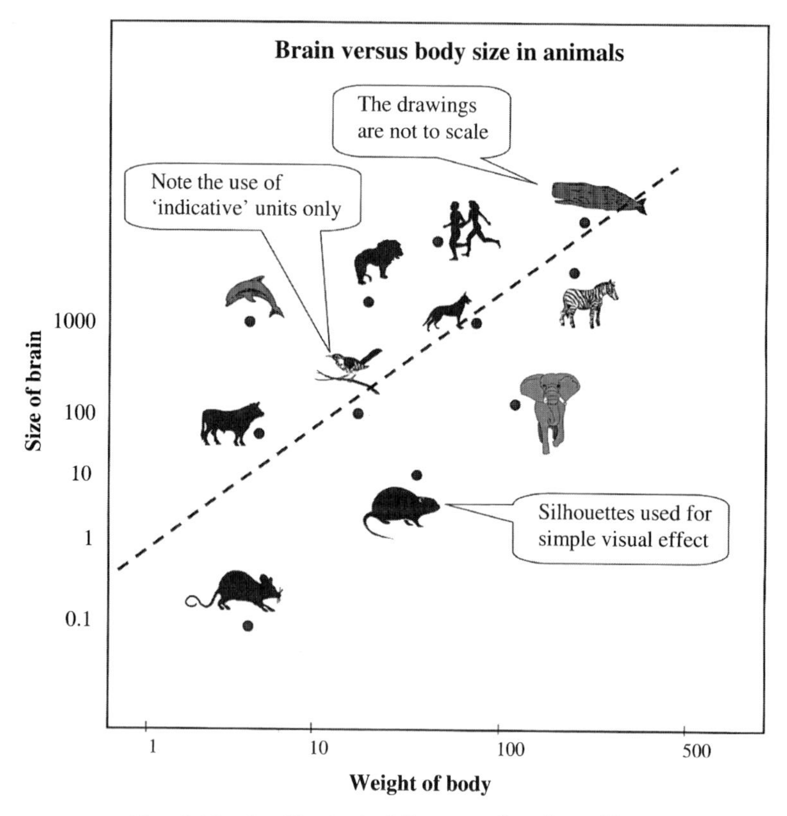

Fig. 2.15 An illustrated line graph using silhouettes

Variations and distortions

- Using incorrect pictures to convey a misleading message. Look at Fig. 2.15 for example. Which is the larger: a mouse or a killer whale?

ILLUSTRATED LINE GRAPHS – TECHNICAL TIP

- Do not use too complex or large pictures in the body of the chart. Use the pictures to help the readers' understanding of the message behind the line curve, not obscure it.

Two-dimensional area charts

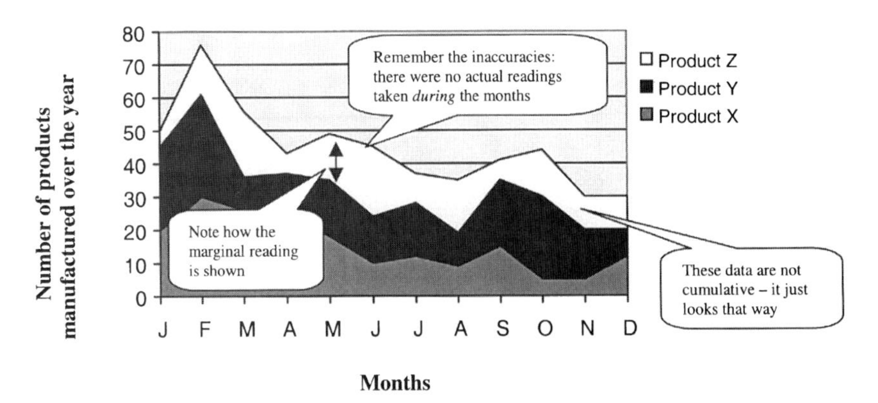

Fig. 2.16 An example of a 2-D area chart

Uses

These are a way of showing trends and changes for either discrete or continuous-type data.

Particularly useful for:

- applications where the objective is to show a *marginal* contribution, i.e. sales, cost, speed, etc.

Data and units

Discrete units of time are shown on the horizontal axis, which is subdivided for accuracy. Colouring or hatching of the areas under the curves is common, to distinguish them from a simple line graph. The inference is, therefore, that the areas under the curves have some significance in the interpretation of the data.

Variations and distortions

As with the previous charts (see Fig. 2.6) the main distortion is the way that discrete data (on the horizontal axis) are sometimes displayed as if they were continuous. Figure 2.16 shows an example of this. Note that it is strictly not possible to interpolate between actual data points, although this is what is inferred.

The most common distortion in interpretation is the way that the areas under the curves appear to mean something, because they are coloured or hatched. This is not necessarily true; you have to look at what the data are actually telling you. Treat each case on its merits.

TWO-DIMENSIONAL AREA CHARTS – TECHNICAL TIP

- Always use solid lines of the same thickness. If you use lines of varying thickness, or dotted lines, as well as colouring/hatching it will cause confusion.

Scatter charts

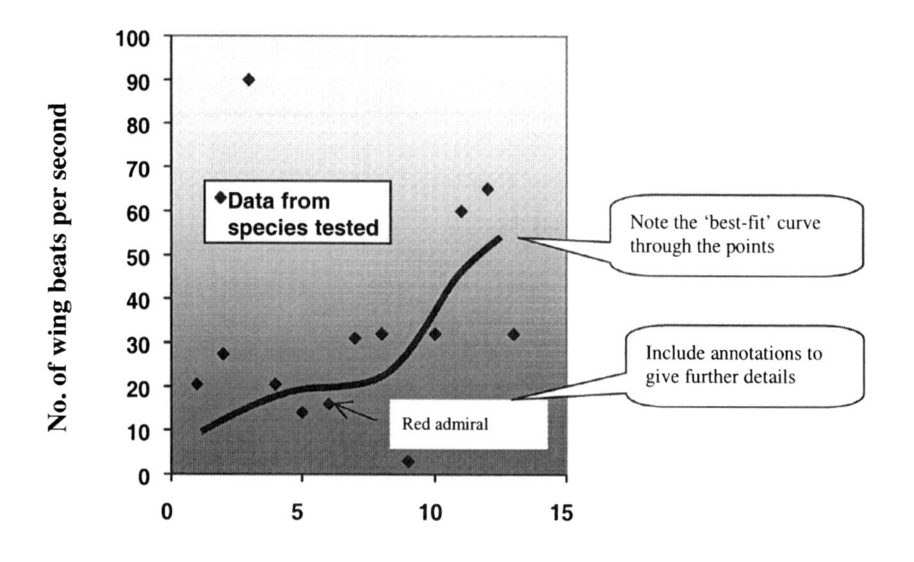

Fig. 2.17 A typical scatter chart

Uses

You may also see these called 'scatter plots' or 'scatter graphs'. They are used mainly to display the relationships between two types of data. Their general objective is to show patterns in the data, so they provide an introduction to the statistical ideas of regression and correlation introduced later in Chapter 10.

Particularly suitable for:

- paired (bivariate) design data in just about all technical disciplines and the natural world;
- data sampled from large 'populations', e.g. mass production, economics, medical statistics, and metallurgy.

Data and units

The axes' scales normally contain continuous data. Several data 'species' may be shown on a single chart. The objective is to draw the best possible curve in relation to the scattered data points that *are* shown. Contrast this with the normal line graph, which may also have been compiled from scattered data, where the data points are not usually shown.

Variation and distortions

- Note how (as in Fig. 2.17) data points can be annotated to provide further 'qualifying' information about the data series they represent.
- The best curve only passes through *some* of the data points. This is not necessarily a distortion; more often the result of computerized curve fitting to robust mathematical rules. It does not necessarily mean the shape of the fitted curve is incorrect.

SCATTER CHARTS – TECHNICAL TIPS

- Never use more than three fitted curves on a scatter chart; it will become too confusing.
- Do not shade or hatch any areas; it is the shape of the curves that is the important display point, not the areas underneath or bounded by them.

Useful reference

http://web.usfca.edn/~villegas/classes/984-

Spider maps

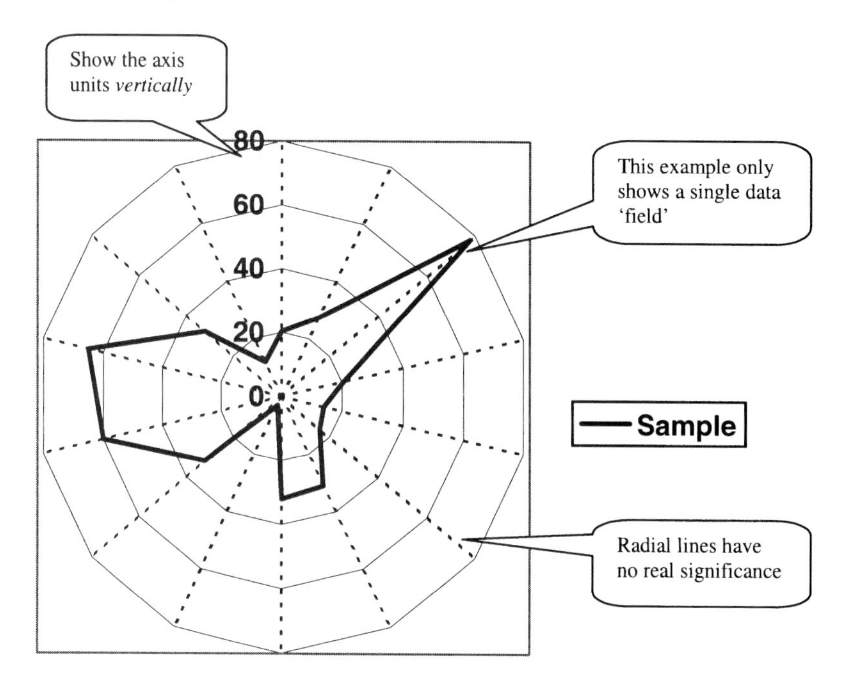

Fig. 2.18 An example of a 'spider map'

Uses

Spider maps are simply a different-looking way to display information that could otherwise be shown in a standard bar graph or line chart. They have few dedicated uses. Their only real purpose seems to be to introduce visual variety.

Data and units

Instead of being plotted on a normal vertical scale, the data points are arranged in a circular form around a central hub. The readings are spaced equally around the 360 degrees, with the radial axis playing the same role as the vertical axis in a standard bar graph. The data may be in continuous or discrete form. The radial points are joined, normally by straight lines.

Variations and distortions

Remember that the location of the data points around the circle has no directional significance. Any annotations that suggest they do can be misleading.

SPIDER MAPS – TECHNICAL TIPS

- Do not make radial lines in the 'spiders web' (as in Fig. 2.18) too heavy, otherwise they will look like they have some significance. They are for appearance only.
- Annotate the radial scale in the vertical direction; if you put it horizontally it might be mistaken for a time scale.

Bubble charts

Uses

Bubble charts have two different applications. In one form they can be used to display separate 'fields' within a graph using standard quantitative horizontal and vertical axes (Fig. 2.19). By far the most popular application, however, is to use bubble charts to convey conceptual information where the axes have no quantitative interpretation. Figure 2.20 shows a typical application.

Data and units

If quantitative scales are used, they invariably contain continuous rather than discrete data. This allows 'circular fields' to be drawn, indicating areas of the data set that meet some specific criteria.

Variations and distortions

The quantitative bubble chart is susceptible to the most common distortion, that of comparisons between the size of the bubbles. Comparisons can be made by either diameter or area, and many charts forget to show which has been used. This can be misleading.

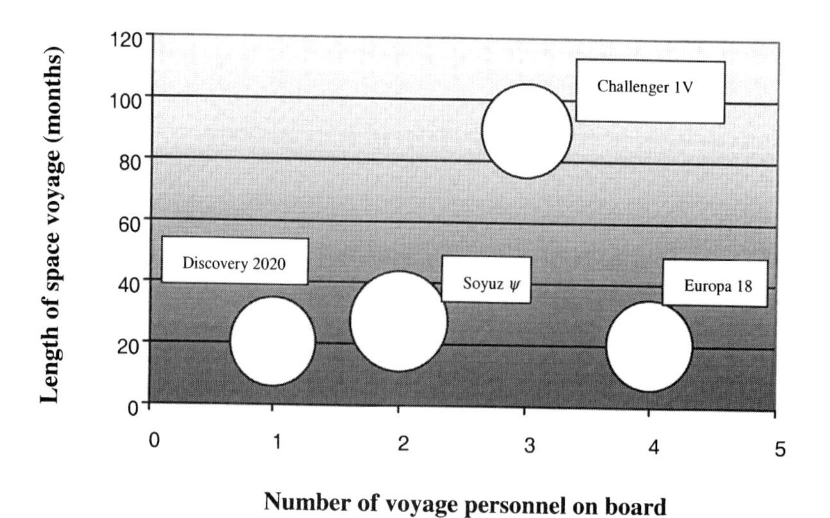

Fig. 2.19 The 'quantitative' type of bubble chart

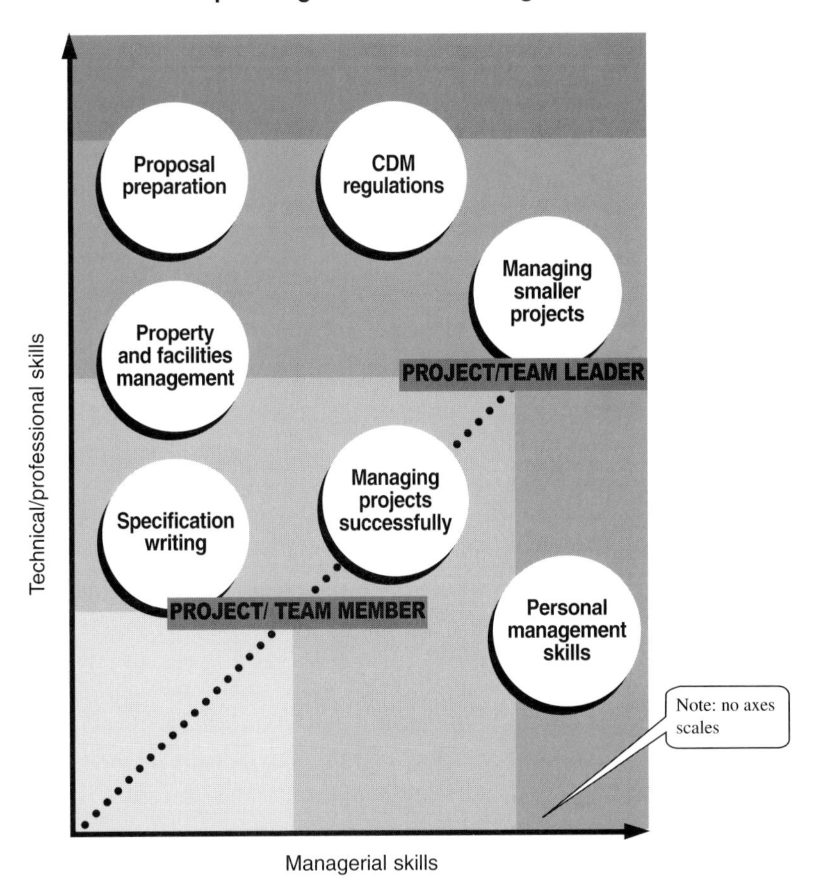

Fig. 2.20 The 'conceptual' type of bubble chart

BUBBLE CHARTS – TECHNICAL TIPS

- 'Conceptual-type' bubble charts must have their bubbles labelled, otherwise they have little meaning.
- Quantitative-type charts can have various sizes of bubbles, but in the conceptual type they are normally the same size.

Useful reference

http://www.julieryan.com/services.htm

Pie charts

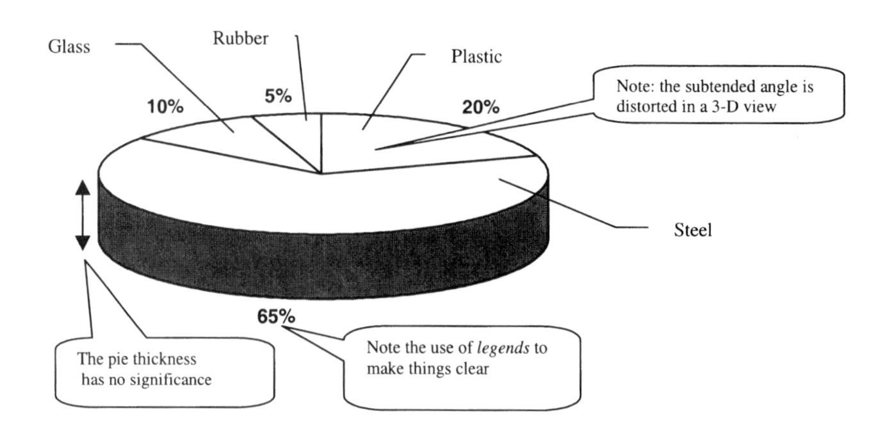

Fig. 2.21 A simple pie chart

Uses

Pie charts show the component parts that go to make up a particular 'set' of technical data. They can be presented in 2-D or 3-D form and generally contain legends, showing the content of each part of the pie (see Fig. 2.21).

Data and units

To be capable of being presented in a pie chart, the data must fulfil two criteria:

- the data are discrete;
- the data set has significance when taken *as a whole,* e.g. 60 seconds in one minute, all the elements in the periodic table, and so on. This second criterion is of key importance when using pie charts; if the full data set has no meaning as a whole, then the breakdown becomes meaningless.

Variations and distortions

- In 2-D pie charts it is the central subtended angle of the pie sections that is the controlling factor in the subdivision. This can become distorted if a 3-D presentation is used. This means 3-D pie charts can be difficult to assimilate accurately.
- Watch out for data included twice.

PIE CHARTS – TECHNICAL TIPS

- Do not use the *thickness* of the pie as a further dimension 'axis'. It is possible, but the result will be open to misinterpretation.

Three-dimensional surface charts

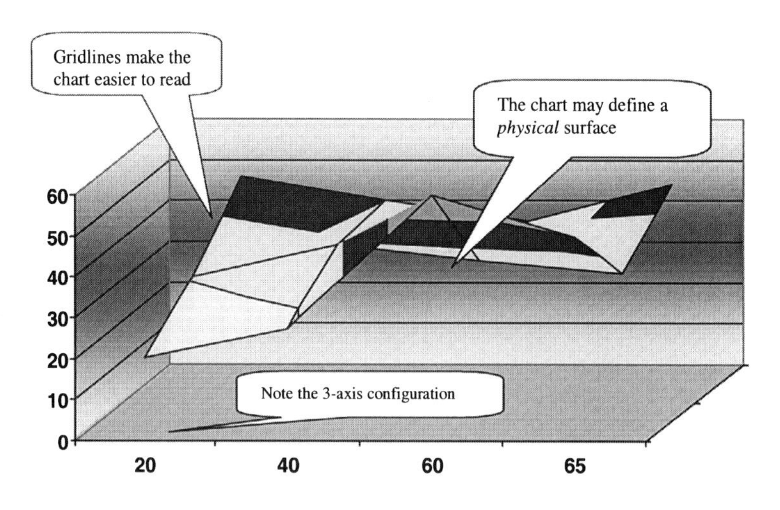

Fig. 2.22 A 3-D surface chart

Uses

Used for either:

- defining a surface (physically) i.e. showing *topology*; or
- showing the planar bounds (edges) of a set of data readings.

Most 3-D surface charts use the three spatial axes x, y, z and refer to physical dimensions in space. There are a few applications where spatial axes are not used, but these are limited to specialist mathematics.

Data and units

Three-dimensional surface charts are difficult to read accurately, but are good at giving a general picture in three dimensions. Expect axes to be only broadly subdivided. Colours or hatching are frequently used to show the surface location on one axis only (usually the vertical one, as in Fig. 2.22).

Variations and distortions

The variation is very wide, ranging from depicting simple geometrical shapes through to concentric spheres, toroids, etc. Most complex ones are computer generated.

THREE-DIMENSIONAL SURFACE CHARTS – TECHNICAL TIPS

- Carefully chosen gridlines on the vertical axis make the chart easier to interpret.
- Remember that the main purpose of these charts is to provide a *general* picture, so do not include too much unnecessary detail.

Combined data charts

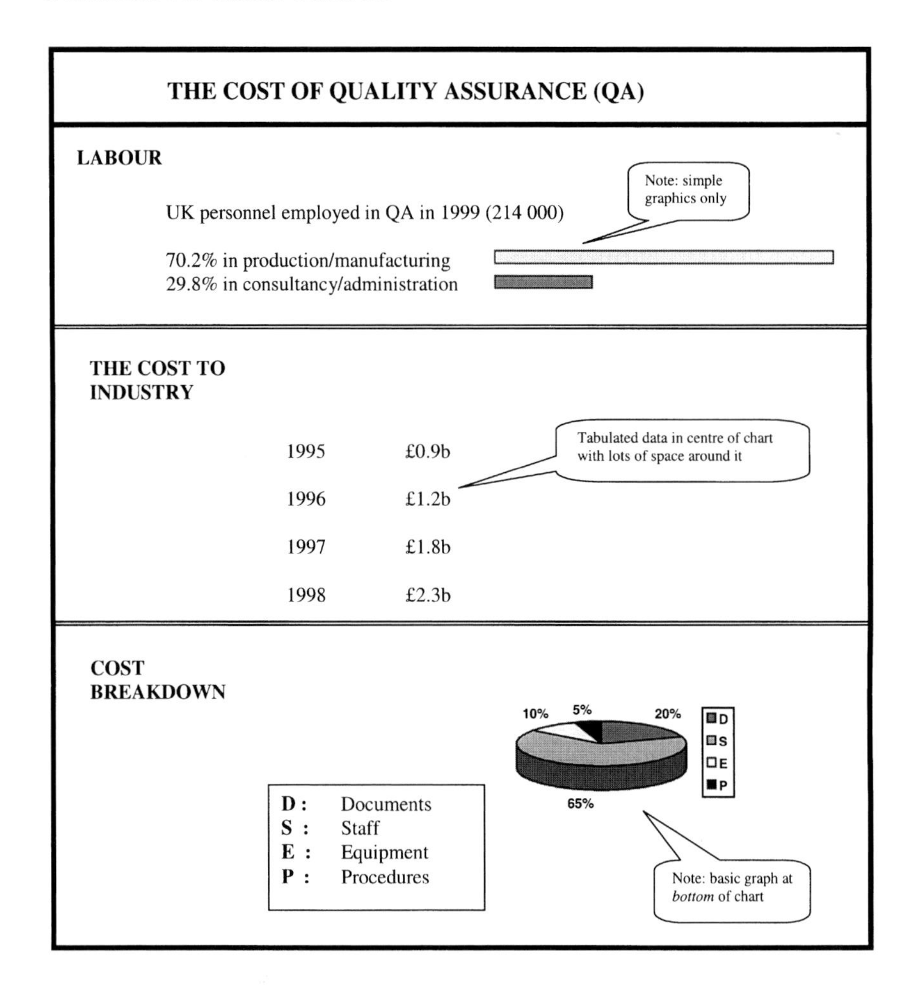

Fig. 2.23 A 'combined data' chart

Uses

Combined data charts are used when it is necessary to show multiple types of information about a particular subject. They are often used as stand-alone technical summaries.

Particularly useful for:

- published articles in technical journals and magazines;
- low-level technical information for presentation to a wide or largely non-technical readership;
- advertising brochures.

Data and units

These charts generally contain three sets of information: a graph, a simple line-scale or trend figure, and some basic figures in percentage or statistical form (see Fig. 2.23). Notice how the information given in each part is different but complementary, helping to build up the full picture. In many cases the information could be combined into a single bar graph or pie chart, but it would need so many subdivisions that it would be almost impossible to read. This is the advantage of combined data charts; they keep things simple.

Variations and distortions

The main variations are in the type of graph that is used. Pie charts fit well with other technical information that is expressed in terms of percentages, while more statistically based situations tend to use bar graphs or histograms. Virtually any form of graph can be used, as long as it is suitable for the context.

Be alert for the following distortions:

- mixing discrete and continuous data on the same chart. This is not *necessarily* wrong (the data may be perfectly accurate and significant), but it can be confusing.
- all of the information is not about the same subject, but presented to look as if it is. This is surprisingly common in financial, demographic, and economic disciplines.

COMBINED DATA CHARTS – TECHNICAL TIPS

- Position the graph at the *bottom* of a combined data chart – it helps with the assimilation of the other information placed above it.
- Position any percentage figures at the *top* of the chart, but keep them simple; two or three at the most.

Pictograms

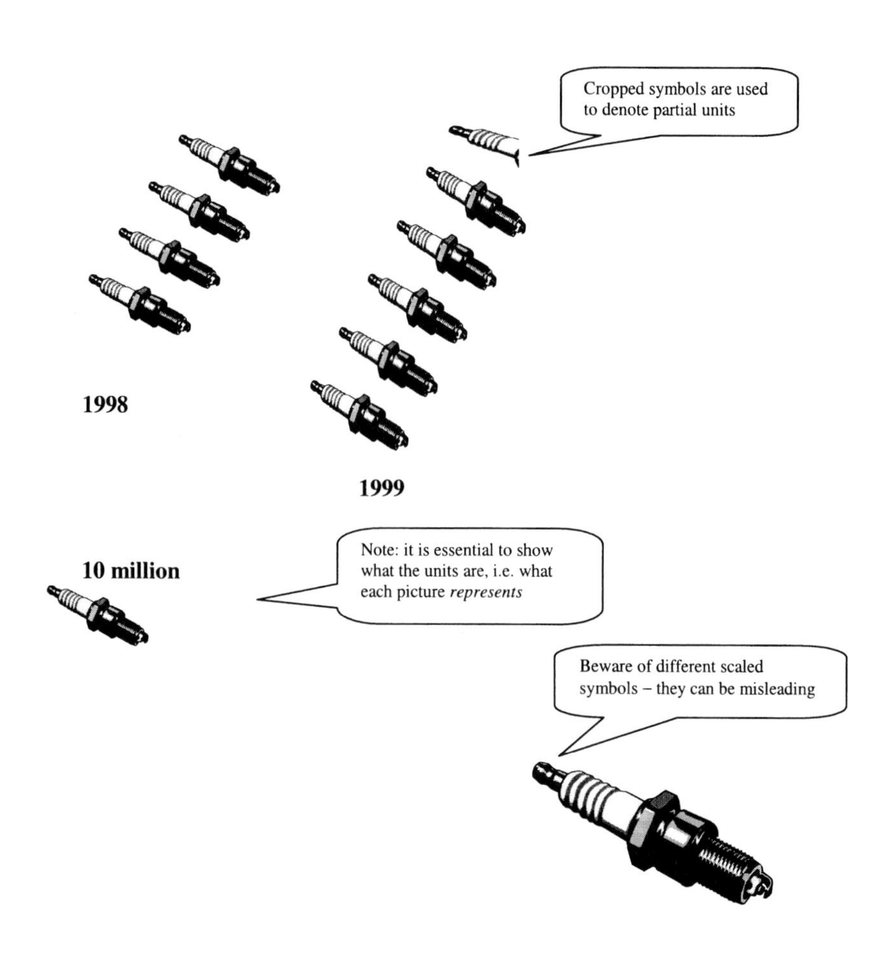

Fig. 2.24 How to use pictograms

Uses

Pictograms, sometimes called pictographs or ideographs, are a pictorial way to represent numbers or frequency of various types of data. They are a very 'visual' method and are easy to assimilate (see Fig. 2.24).

Particularly suitable for:

- displaying information about products that can be shown as a small picture, e.g. technical objects and manufactured products.

Data and units

The pictures are traditionally laid out in straight lines (to partially simulate the effect of a vertical or horizontal bar graph). This makes the data easy to assimilate and compose. A legend is needed to show what unit each picture actually represents. Note how partial units can represented by 'sawn off' pictures. This is not a very accurate method, and the quality of the data presentation is often relegated to being used for comparisons, or 'for guidance only'.

Variations and distortions

The worst example is:

- making some symbols different sizes without explanation. This makes it difficult to know the relative unit value of each symbol, so the information becomes almost meaningless.

PICTOGRAMS – TECHNICAL TIPS

- Horizontal lines of pictures are easier to compare than vertical ones – the eye seems to follow them better this way.

Useful references

http://www.pictograms.com/pictonews.html
http://www.pictograms.com/bookrelease.html

Cycle charts

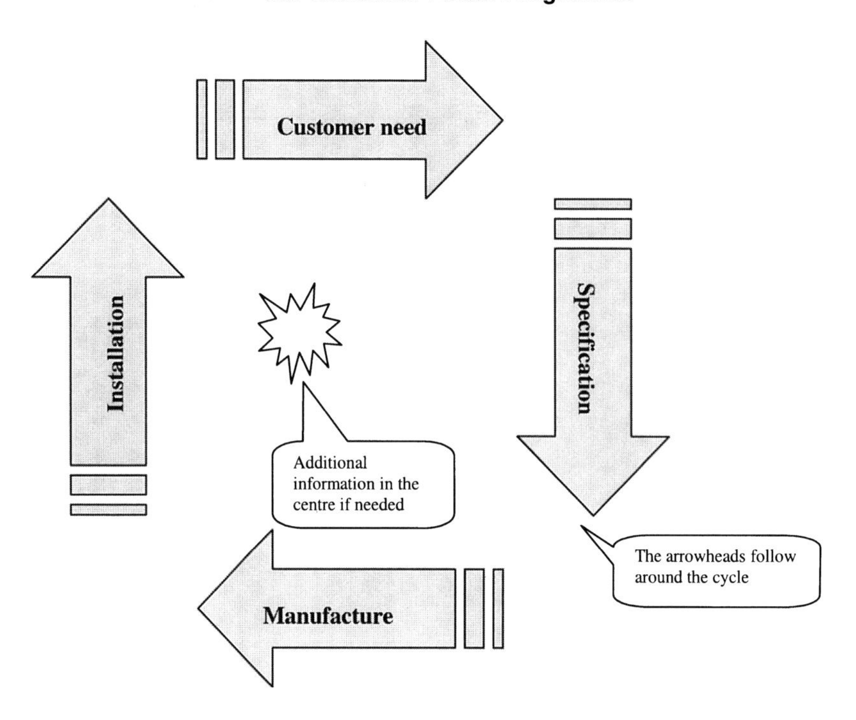

Fig. 2.25 The basic 'cycle chart'

Uses

Cycle charts are used whenever you want to show a process or series of events which are related in *time*. The series of events may be complete in itself or form part of some larger sequence.

Particularly useful for showing:

- manufacturing processes;
- quasi-technical activities such as quality assurance;
- project management stages.

Data and units

There are no units. The steps are arranged, in a circular pattern, in basic chronological order. There is not always a 'start point', but if there is it is normally shown at the 12 o'clock position.

CYCLE CHARTS – TECHNICAL TIP

- Do not forget to use arrowheads of some sort, to make it clear which way the cycle is going.

Combined cycle/context charts

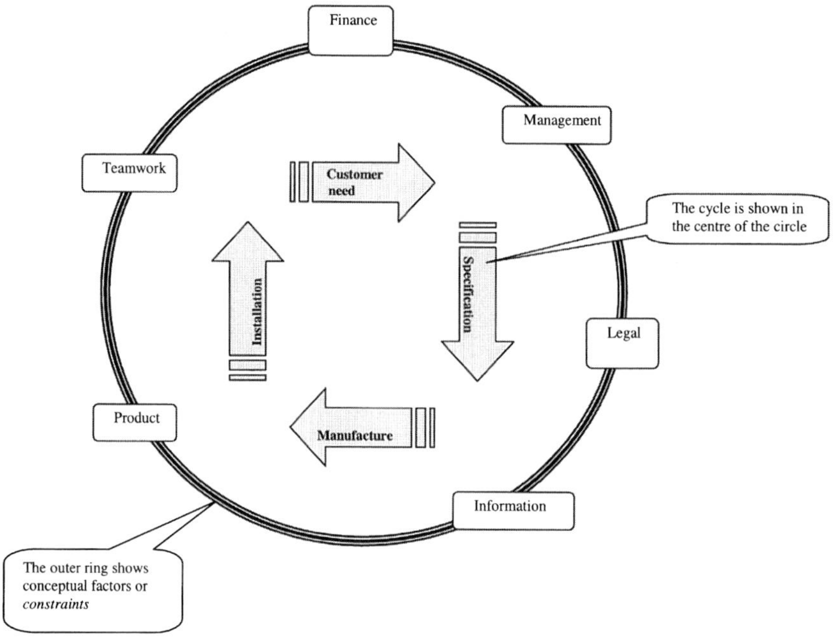

Fig. 2.26 The combined 'cycle and context' chart

Uses

These are a common extension of the simple cycle chart. The centre circle shows a chronological sequence of events, while the outer circle shows the context that surrounds these events (see Fig. 2.26). The context may be technical or procedural and usually acts as some kind of *constraint* on the cycle.

Particularly suitable for:

- visualizing technical management situations;
- any technical *design* discipline involving 'conceptual' steps.

COMBINED CYCLE/CONTEXT CHARTS – TECHNICAL TIPS

- Remember that the outer ring shows context, or conceptual constraints; it is not a cycle, so do not show it as one.
- Do not infer any time links between the inner and outer rings. The outer ring shows *context*, so it is there all the time.

Diametral charts

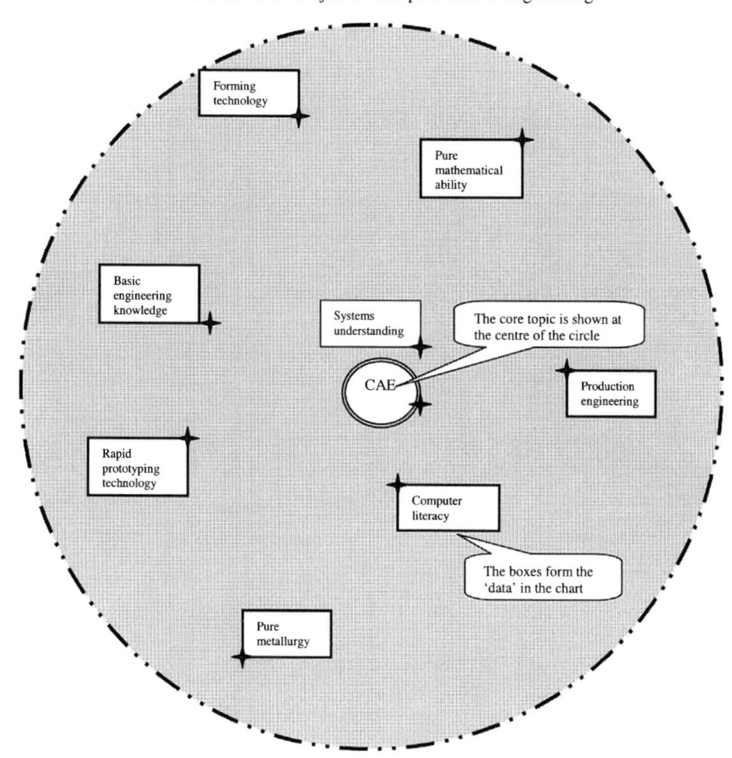

Fig. 2.27 A diametral chart showing 'conceptual distance'

Uses

The purpose of diametrical charts is to show conceptual distance from a particular subject or concept shown at the centre of the chart. Hence, the position of the other subjects or concepts within the main chart circle indicates the closeness of their relationship with the one situated at the centre.

Particularly suitable for showing:

- interrelationships (or lack of them) between technical disciplines;
- 'softer' technical subjects such as control theory, neurology, cybernetics, systems analysis, and organizational design.

Data and units

The contents of the various boxes forms the 'data' in the chart. These data categories are normally discrete from each other (if they were not, they would be shown as overlapping). Note how it is only the radial position of each box that has significance, the circumferential spacing being chosen for convenience and clarity only. No scale is shown, because it is only conceptual distance that is being portrayed; there is no quantitative link with the physical world.

Variations and distortions

Using boxes of different sizes. This is misleading because the purpose of the chart is to show the conceptual *position* of a subject or concept, not its size or complexity.

DIAMETRAL CHARTS – TECHNICAL TIPS

- Use points or stars to locate the position of each box (as in Fig. 2.27); it makes things clearer.
- Always use a broken or dotted line to enclose the 'conceptual circle', because everything in the conceptual world has flexible boundaries.

And then: graphical imagination – the link with calculus

How do you show information about things that change? It is certainly necessary; all technical subjects (and the rest of the world as well) are full of motion, flow, cycles, and patterns. Everything in fact, is in a state of change, even if it does not look like it.

One way is to use graphs. We have seen that various types can be used to present information in the way that we want.

So what is the problem?

The problem is that graphs cannot tell you everything you need to know. They do show things that are changing but they are based only on analytical geometry and have trouble, for example, in defining precisely *the rate* at which things are changing. Look at the graph in Fig. 2.28. It certainly would not be possible from this to measure accurately the length of the curve, or the area underneath the curve. How would you do this?

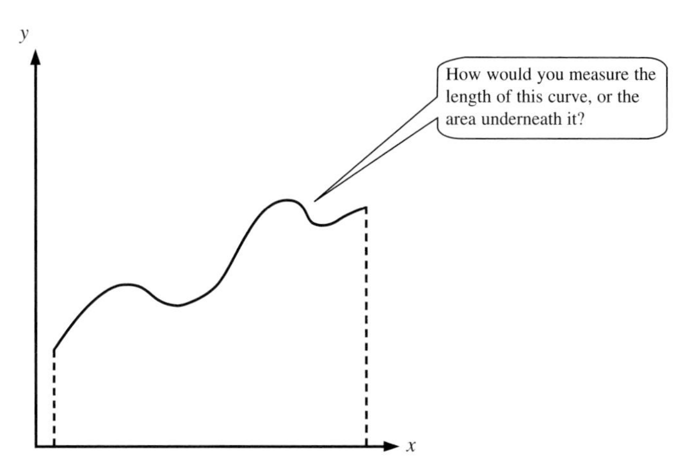

Fig. 2.28 The problem with graphs

The start of the answer

One way to start to address the problem is to think of the graph (all graphs) as a series of static pictures. Divide the curve (and the area underneath it) into small bits, and then very small bits; try one thousand per centimetre. Now keep going until you have a million bits slotted into each pin-prick of curve length, each with its rectangle of width one millionth of a pin-prick hanging underneath it. Fig. 2.29 shows the result.

Now the key points:

- if you make the number of subdivisions infinitely large then:
  - the sum of the top of the rectangles equals the length of the curve;
  - the areas of the rectangles add up to the area under the curve.

Revelation

This transforms our 'graph' into a tool for expressing *calculus* – the subject of change and rate of change. Calculus involves the differential and the integral – concepts and rules related to rates of change and its accompanying sums and areas. These rules extend the usefulness of our previously harmless graph into a technique that can deal with the natural laws that underlie all technical subjects. We now have a tool that we can apply to phenomena such as falling bodies and things that undergo growth or decay. Calculus is the language of the natural world (and so the technical one as well).

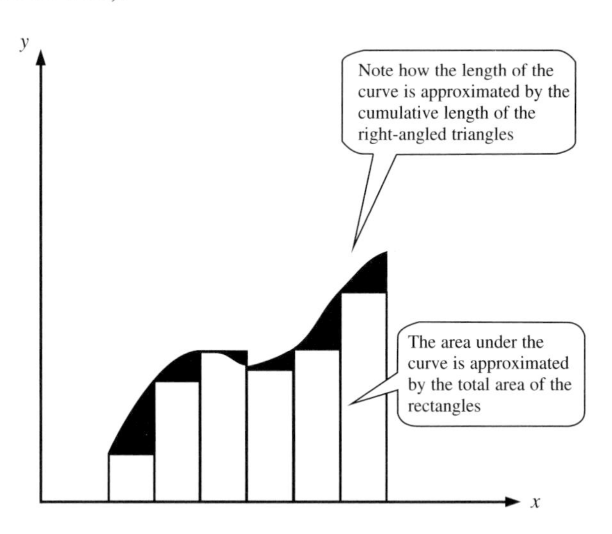

Fig. 2.29 The idea behind calculus

Chapter 3

Technical Drawings

Introduction: the role of the technical drawing

Drawing is a graphical language. It is needed because accurate technical information cannot be conveyed in words. The fundamental purpose of technical drawing is to convey the exact shape and dimensions of an object, i.e. provide unique information about its geometry.

> **TO CONVEY ACCURATE TECHNICAL INFORMATION:**
>
> DRAWINGS ARE *ESSENTIAL*, NOT OPTIONAL

Drawings exist as sets

This is an important concept. Even the simplest technical design or object may need several different types of drawings to provide full information about it, while a complex product – such as a motor vehicle or airliner – will need many thousands.

Drawings use conventions

Fortunately, the complexity of the vast range of technical drawings is kept in check by the use of proven sets of conventions. These conventions are not simple; there are many different types of special conventions used in design, engineering, architecture, and other forms of technical drawing. They are, however, *consistent* and form an almost universal technical language that does not require the use of words. In this and subsequent chapters, we will look at the more popular conventions, and how they are used.

The use of CAD

Computer Aided Drafting (CAD) is well established in most technical disciplines. It automates the drawing process using proprietary 2-D or 3-D software packages and provides great advantages in the speed and accuracy of drawing. The techniques and conventions are, however, almost exactly the same as those developed for manual drawing, so you can think of CAD as simply an automation of established techniques, rather than a new technique.

General practice and conventions

Lines

In technical drawings, line types have specific meanings. The exact conventions differ only slightly between technical standards. The basic set is shown in Fig. 3.1.

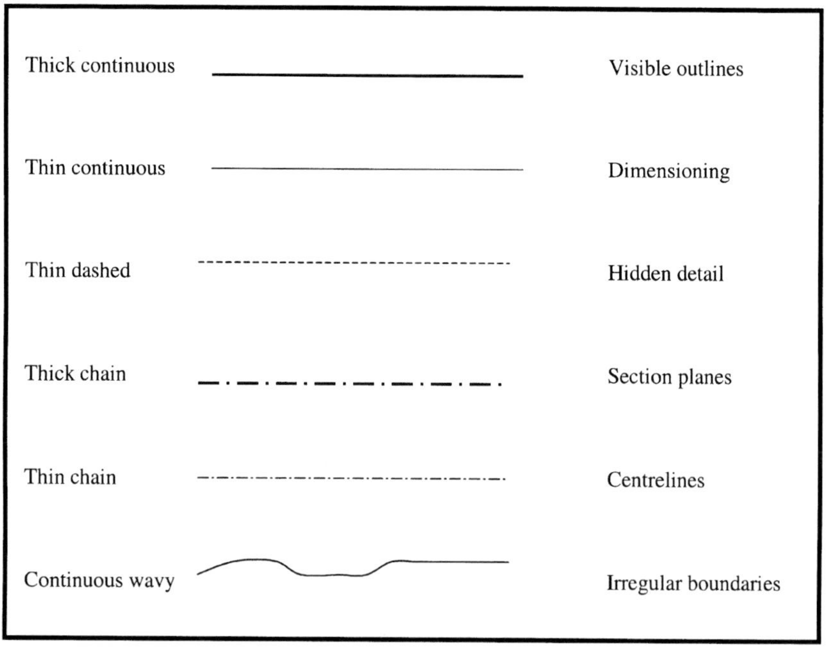

Line type	Use
Thick continuous	Visible outlines
Thin continuous	Dimensioning
Thin dashed	Hidden detail
Thick chain	Section planes
Thin chain	Centrelines
Continuous wavy	Irregular boundaries

Fig. 3.1 Technical drawing convention – line types

Figure 3.2 shows how these line types are used in a simple technical drawing.

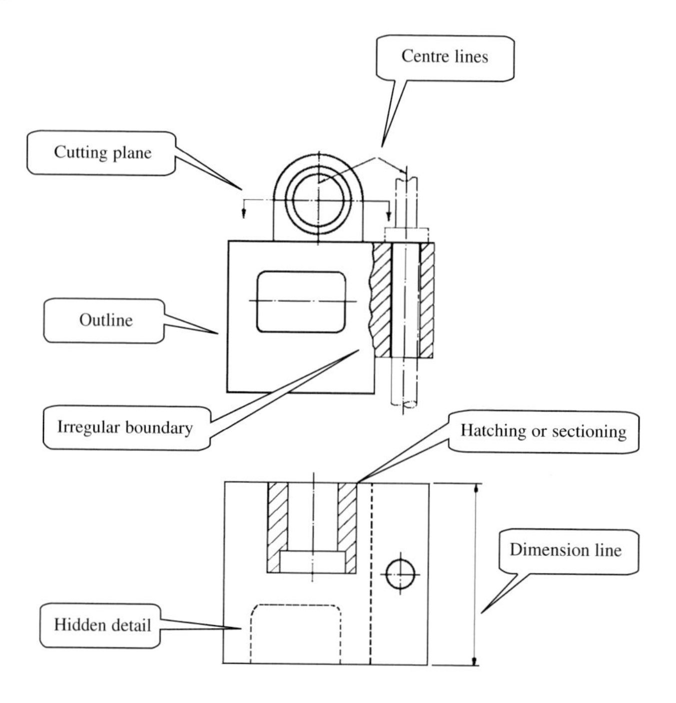

Fig. 3.2 How line types are used

Dimensions

Dimensions are a key part of most types of technical drawing. Conventions apply in two areas:

- how dimensions are shown;
- how dimensions are applied to the drawings to which they relate.

The basic methods are shown in Fig. 3.3(a) and (b).

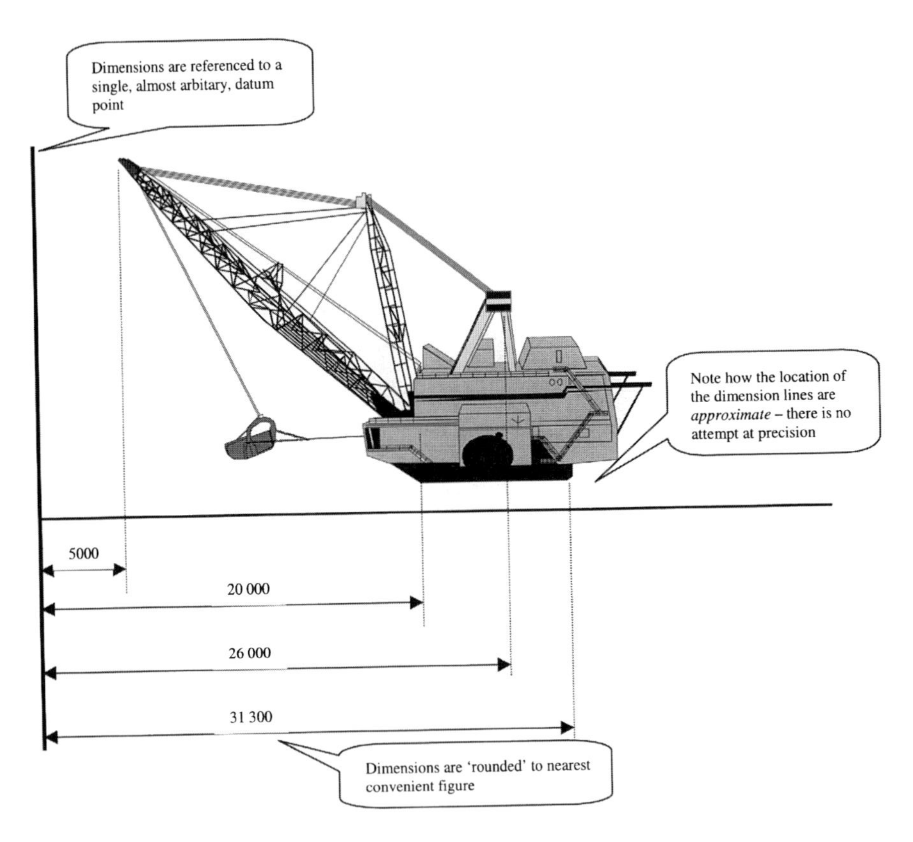

Fig. 3.3(a) 'General' dimensioning of a sketch

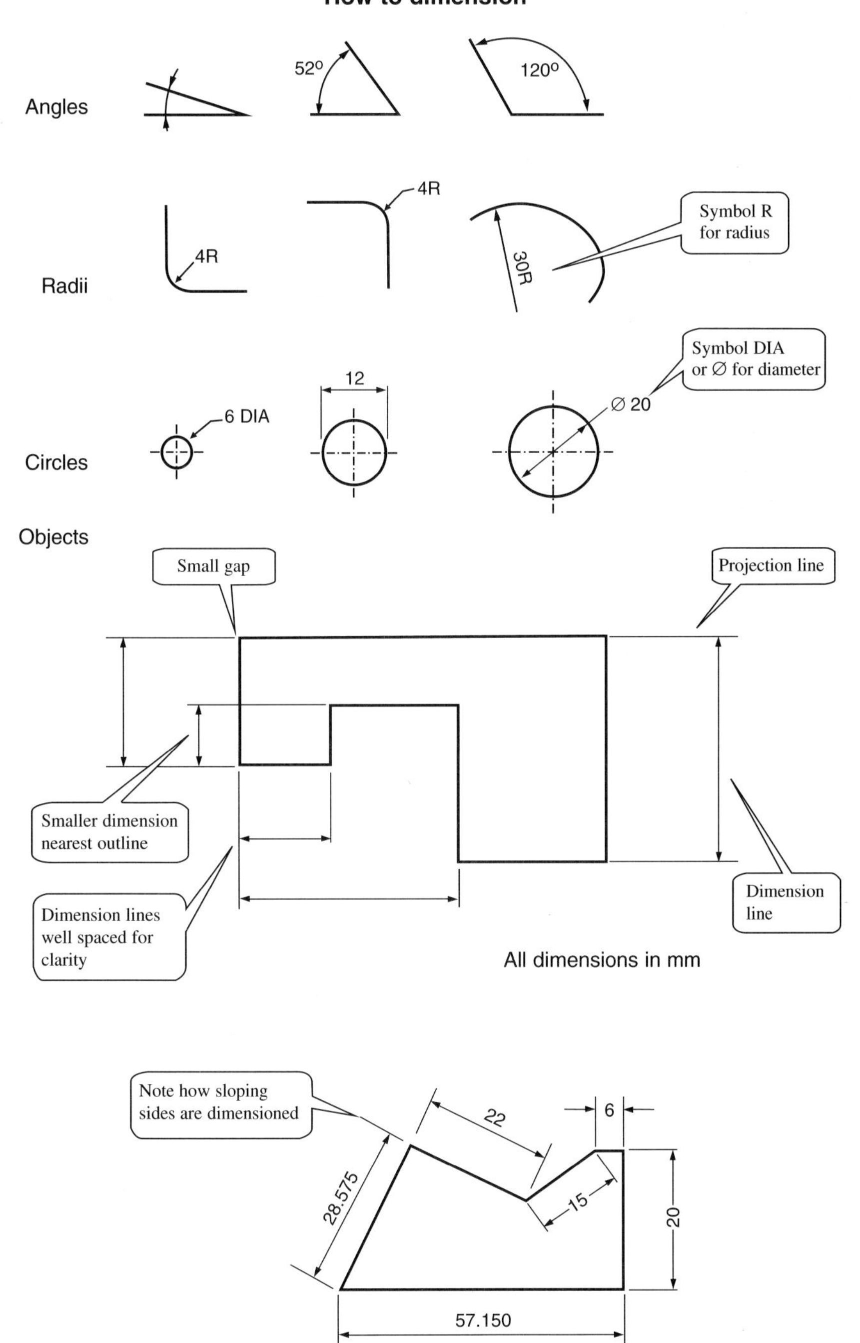

Fig. 3.3(b) Conventions for *precise* dimensioning of technical drawings

Orthographic projection

The idea

Orthographic projections are one of the most common forms of technical drawing. The technique involves representing an object by a set of 2-D views or 'elevations'. In its simplest form, one or more views are shown of a technical object, without accurate dimensions and perhaps including some explanatory text. Figure 3.4 shows an example.

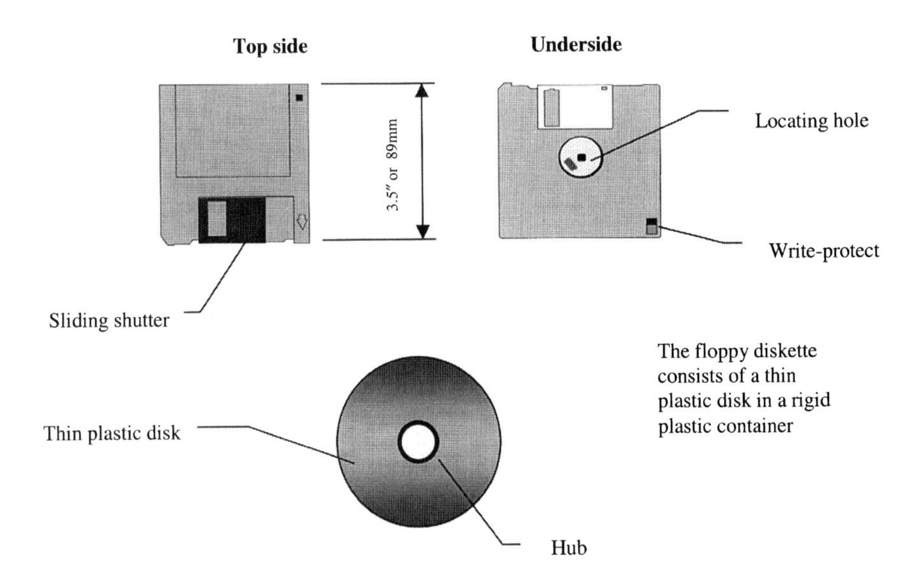

Fig. 3.4 A basic form of orthographic projection

Orthographic projection – what is it?

Proper orthographic projection (not the loose version of Fig. 3.4) is a rigid, conventional system of drawing that is capable of depicting the precise geometry, dimensions, and features of an object. It does this using a set of separate 2-D views or 'projections'. Each view is not complete in itself, but when taken with others, the set provides a full description of the detail of the object.

> **THE MAIN ORTHOGRAPHIC VIEWS ARE: (SEE FIGS 3.5 AND 3.6)**
>
> - The front elevation (FE) – this usually shows the side of the object which has the longest dimensions.
> - The plan – a view from above the object.
> - Side or end elevation(s) – view(s) from the end of the object.

Orthographic drawings can be presented in one of two ways: first angle projection or third angle projection (see Figs 3.5 and 3.6). These are simply conventions, i.e. different ways of doing the same thing. Note the symbols on the drawings, showing which projection has been used.

First angle projection

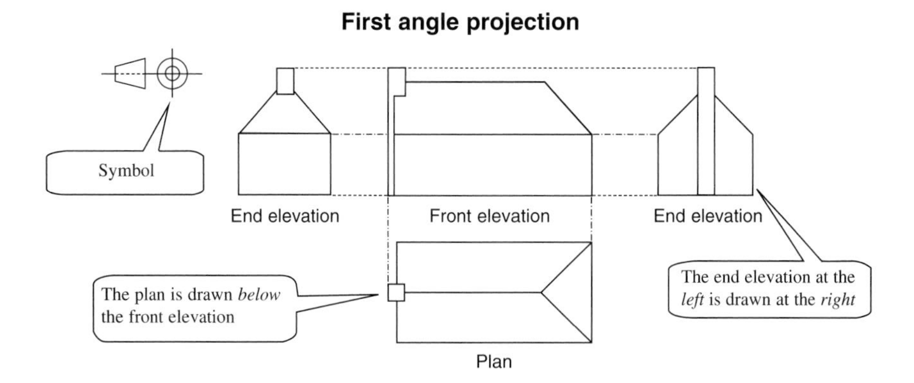

Third angle projection

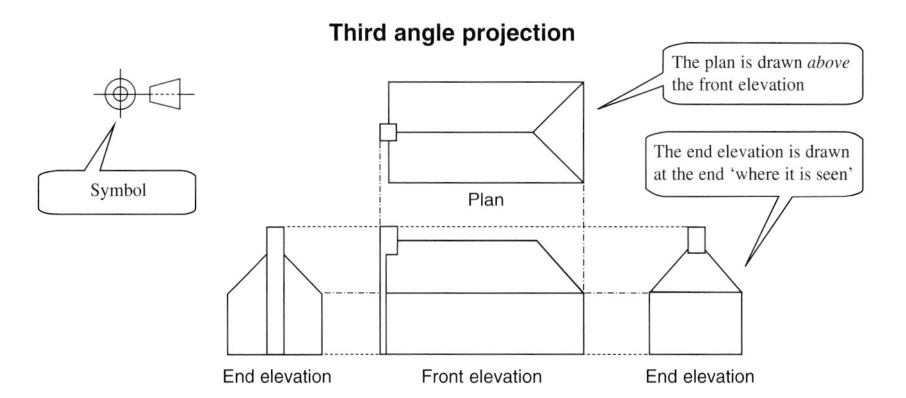

Fig. 3.5 First and third angle orthographic views

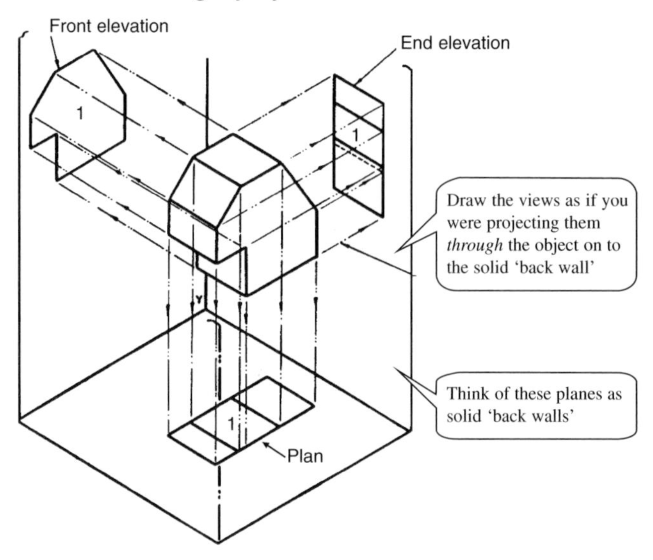

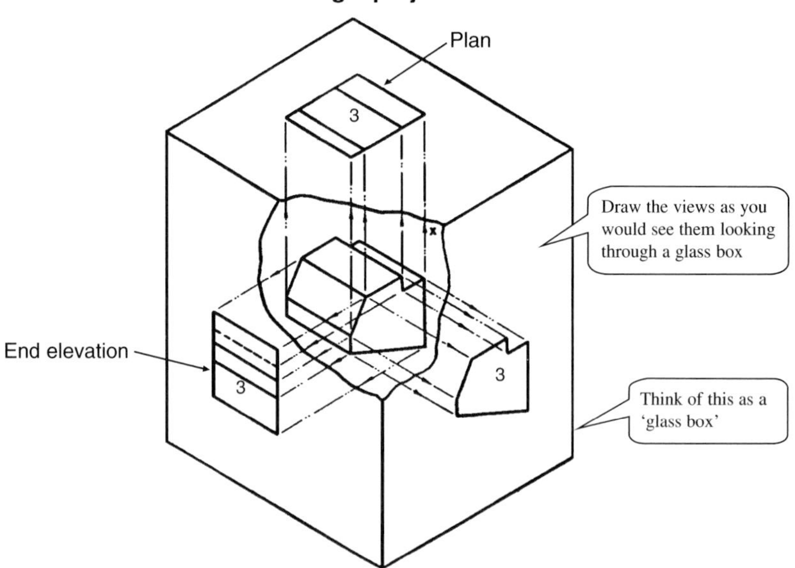

Fig. 3.6 First and third angle views – another way to think of them

Useful references

http://machinetool.tstc.edu/CNCC2.HTML
http://www.foothillsgraphics.com/ortho.htm

Sectional views

Sectional views or *'sections'* show an object as if part of it has been cut away. The purpose of this is to show internal detail when this is too complicated to show using 'hidden detail' lines on a normal orthographic external view. Hidden detail lines are, therefore, not used on sectional views. There are several different types of sectional view, depending on what you want to show (see Figs 3.7 and 3.8), and also a set of conventions to be used with them.

Full section

Here the imagined cutting plane passes completely through the object. Cut surfaces are shown as hatched. It is convention not to hatch shafts, webs, bolts, and fasteners (see BS 308 **(1)** for more details).

Half-section

Half-section views are useful for symmetrical objects (Fig. 3.7). One half is drawn as a section with the other left as a normal outside view. Both halves are then shown joined together to save space. Their main use is in assembly drawings (showing how objects fit together).

Scrap section

This is used to show a few necessary details of the construction of an object without resorting to a full-sectional view. Note how the sectioned part can be moved out of the main view for clarity (Fig. 3.8).

Revolved section

This is used when it is necessary to show the cross-sectional shape of an object. It can be either drawn on the outside view or in-between 'broken' bits of it, as shown in Fig. 3.8.

Full section view

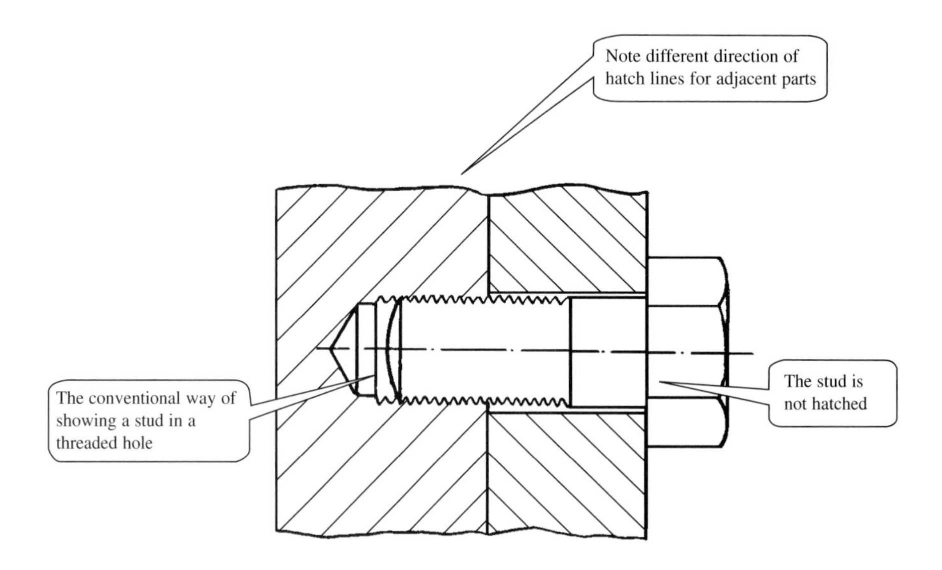

Half-section view

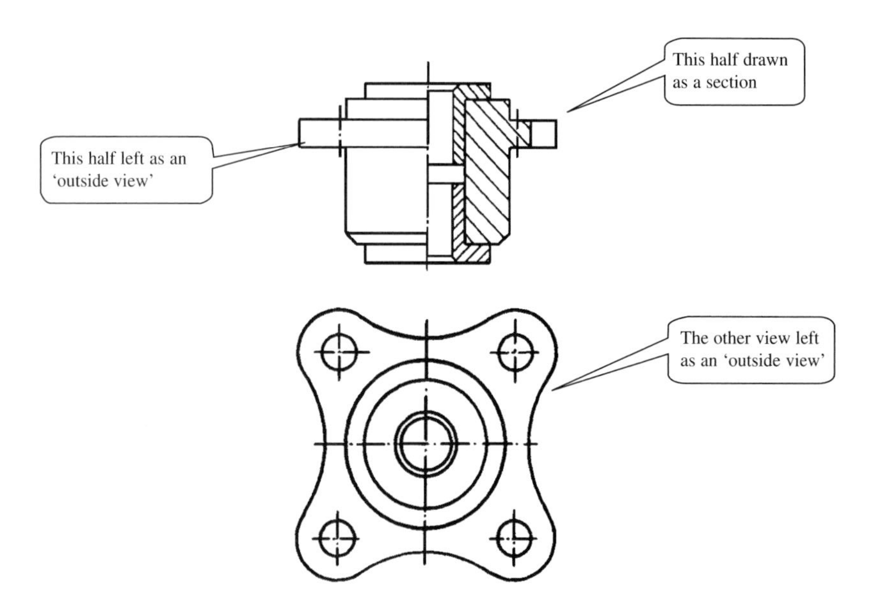

Fig. 3.7 Some sectional views

Scrap section views

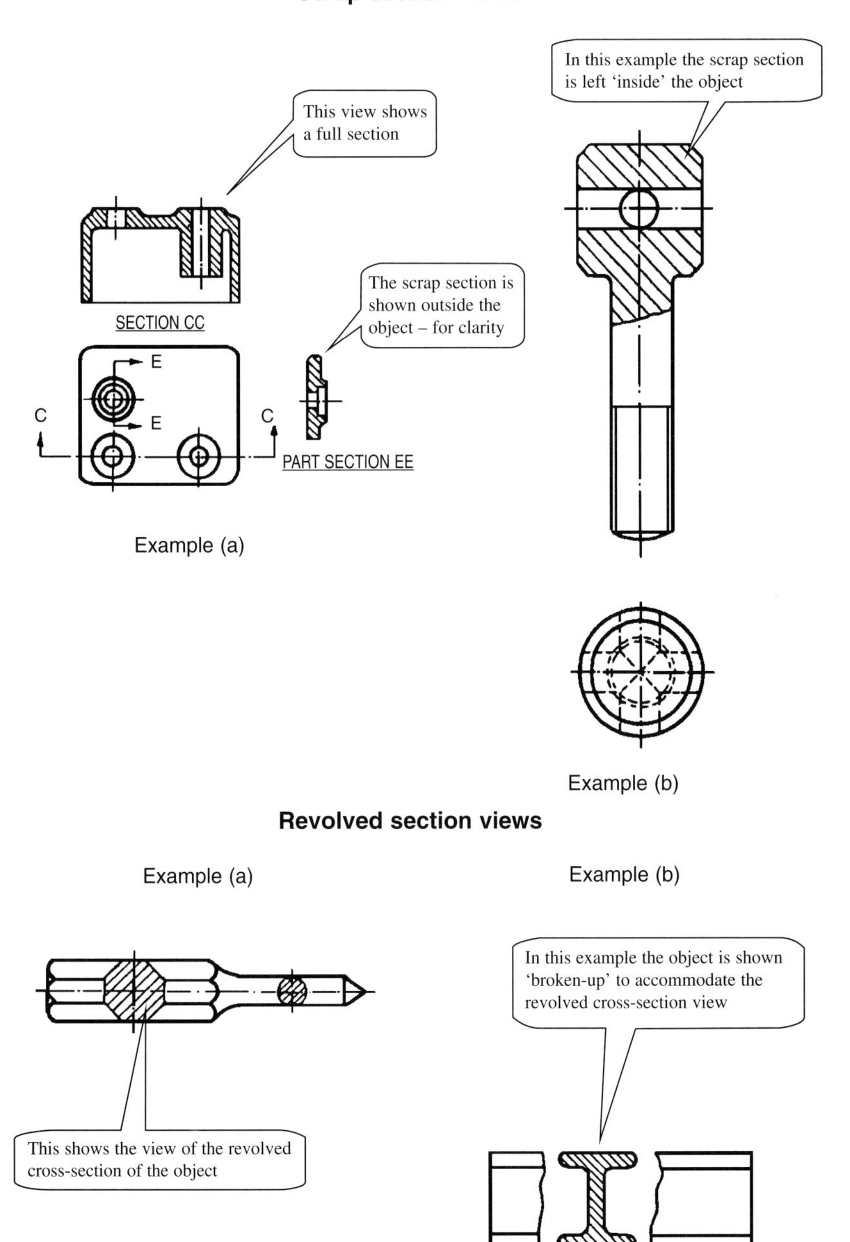

Fig. 3.8 More sectional views

Isometric projection

Isometric projection is the most common way of showing a 3-D 'pictorial' view of an object without the use of perspective (see later). The drawing is built up on axes orientated at 30 degrees to the horizontal. The method does not give an absolutely true 3-D representation because there is no allowance for convergence of lines as they recede into the distance. It does, however, have the advantage that a single scale of measurement can be used, and the views are not foreshortened.

ISOMETRIC DRAWINGS – KEY FEATURES

- The views are distorted, so circles appear as ellipses.
- You cannot easily 'scale' from the dimensions.
- They do not have hidden detail (dotted) lines.

Figure 3.9 shows a typical isometric technical drawing of a simple solid object.

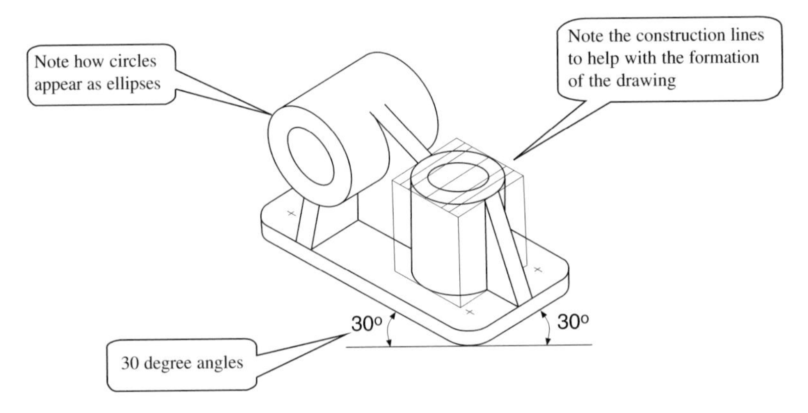

Fig. 3.9 Simple isometric projection

General isometric applications

Isometric views are also used in many discipline applications to provide general technical information about 3-D objects. Figure 3.10 shows typical examples.

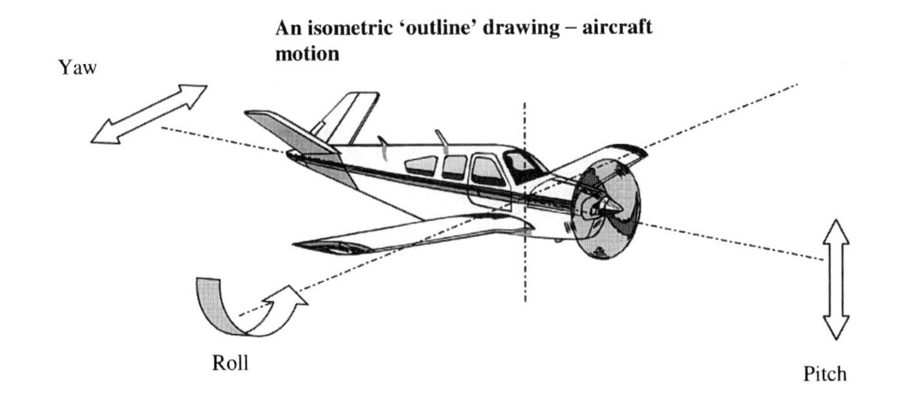

A 'shallow angle' isometric

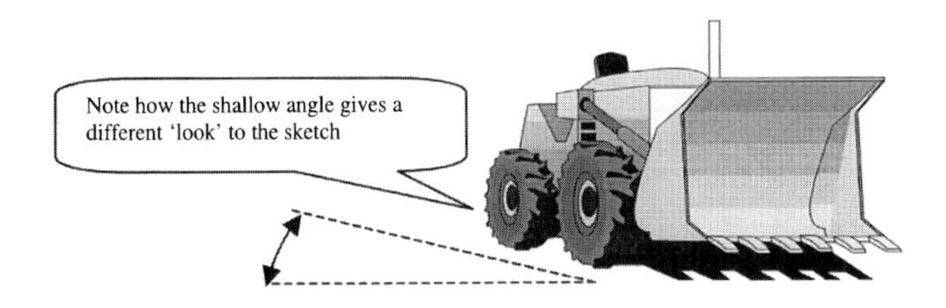

An isometric 'pictorial' (from an instruction manual)

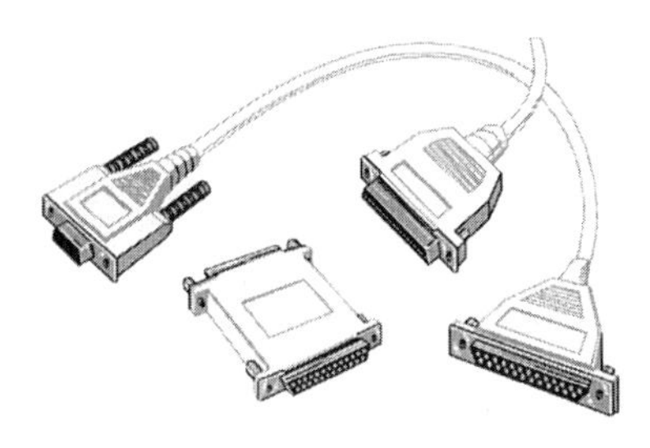

Fig. 3.10 Typical uses of the isometric projection technique

Oblique projection

Oblique projection is an alternative way to show a 3-D pictorial view of an object.

> **OBLIQUE DRAWINGS – KEY FEATURES**
>
> - The front face is drawn as a true (undistorted) view; this is useful if the object is geometrically complex.
> - The 'receding' axis is orientated at 30, 45, or 60 degrees to the horizontal.
> - Dimensions on the receding axis are reduced (normally by 50 percent).

Figure 3.11 shows several different ways in which oblique projection is used.

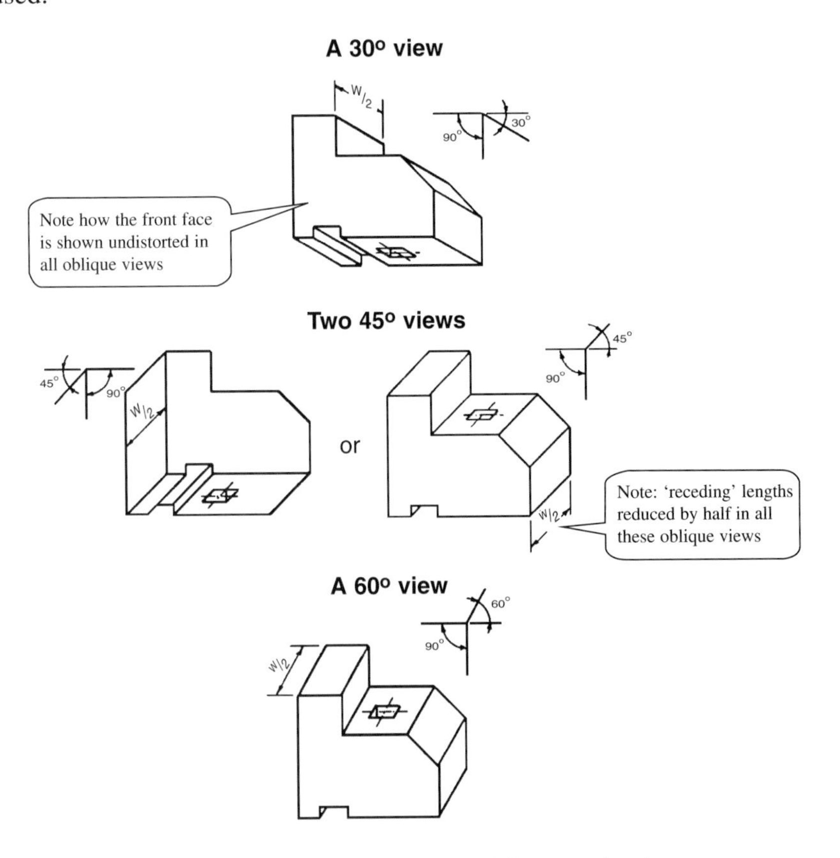

Fig. 3.11 Different techniques of oblique projection

Exploded views

This is the technique of showing an object 'exploded' apart (see Fig. 3.12). Exploded views normally use isometric projection. The axis angle may be reduced from 30 degrees to 10–20 degrees (to give more horizontal space) if the item is complex.

Uses

- Drawings showing component identification (e.g. spares catalogues).
- Instruction/maintenance manuals, showing how things work.

Note how there is no attempt to show accurate technical details, either about the subcomponents or how they fit together, so these types of drawings are mainly for 'guidance only'.

A view exploded in one plane (horizontal)

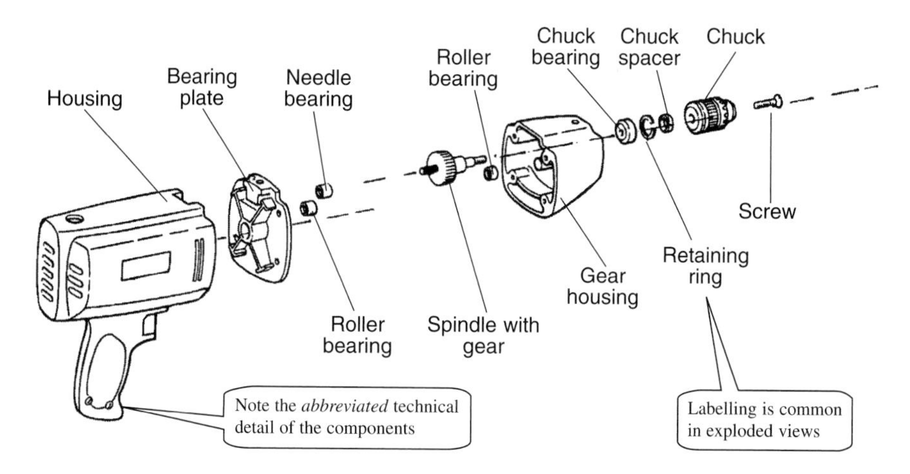

A view exploded in two planes (horizontal and vertical)

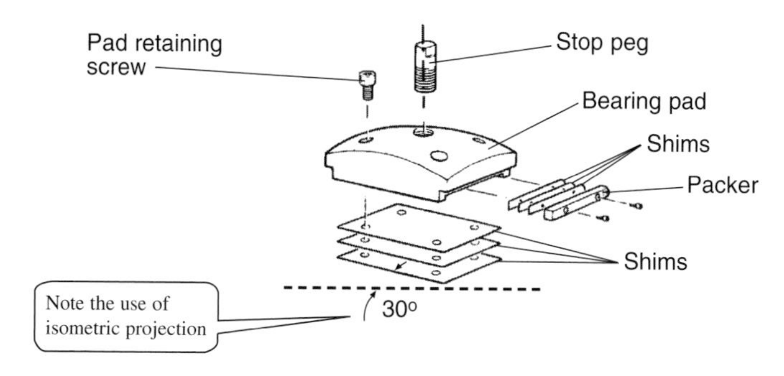

Fig. 3.12 Two types of exploded view

Loci

When a point moves through space, its path is known as a locus (plural *loci*). Loci are an important consideration in the design of mechanisms and machines, in fact anything that contains moving parts. Because of their nature, loci follow strict mathematical rules and so can be represented either by diagrams or mathematical expressions. Some common loci found in technical disciplines are:

- circles;
- ellipses;
- parabolas/hyperbolas;
- helices.

For the purposes of showing loci, we can divide them informally into two types: rigid and non-rigid.

Rigid loci

These are used in accurate technical drawings for the design of engineering mechanisms such as cams and linkages. The shape of the locus is plotted in stages and determined accurately for use in the design process. Figure 3.13 shows an example.

Locus of a wheel pivot as it traverses a track

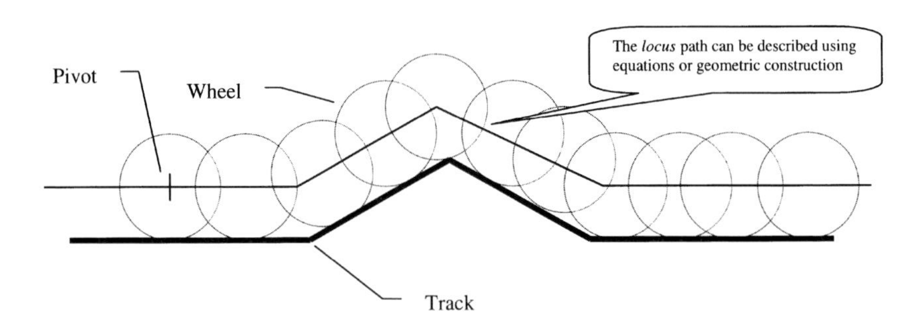

Fig. 3.13 A 'rigid' locus drawing

Non-rigid loci

In this context, the term 'non-rigid' infers that the locus is expressed loosely with the objective of only showing the *general path* of motion of a point or object. Such information is, therefore, 'for guidance only' and cannot be processed in a mathematical equation or used as an accurate input parameter to the design process. Figure 3.14 shows typical examples.

Typical uses of non-rigid loci diagrams are:

- transportation – showing the path of vehicles, aircraft, etc.;
- early design sketches of machines and mechanisms (see Chapters 4 and 5).

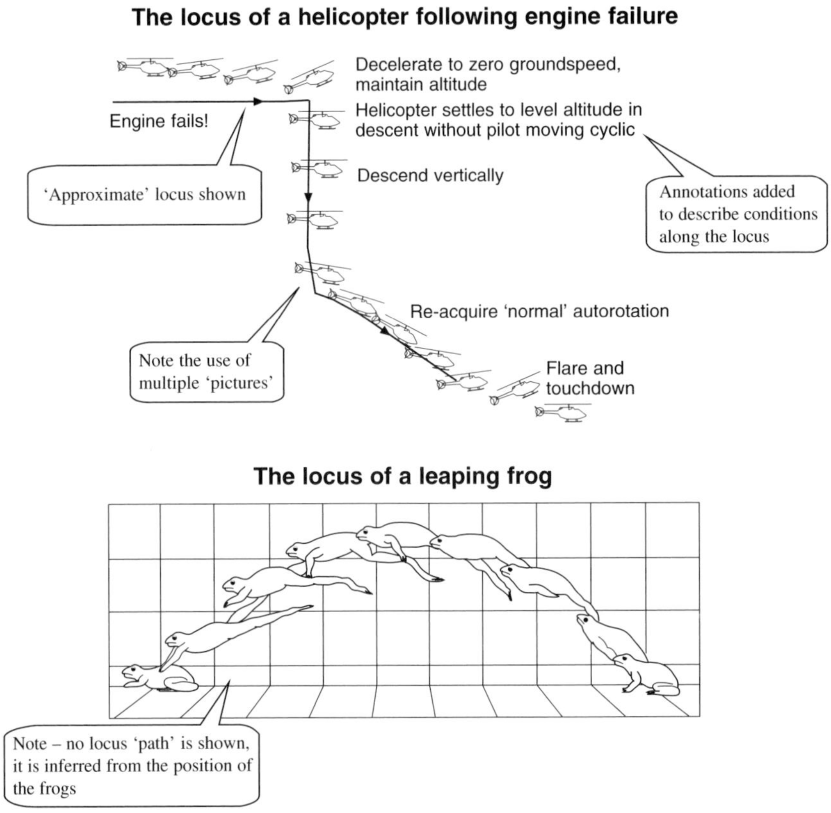

The locus of a helicopter following engine failure

Decelerate to zero groundspeed, maintain altitude
Engine fails!
Helicopter settles to level altitude in descent without pilot moving cyclic
'Approximate' locus shown
Descend vertically
Annotations added to describe conditions along the locus
Re-acquire 'normal' autorotation
Note the use of multiple 'pictures'
Flare and touchdown

The locus of a leaping frog

Note – no locus 'path' is shown, it is inferred from the position of the frogs

Fig. 3.14 Examples of 'non-rigid' loci drawings

Perspective drawings

What are they?

Perspective drawings occupy an uneasy position half-way between a pure technical and a pictorial drawing. Perspective drawings for technical (as opposed to artistic) purposes follow a well-defined set of rules.

THE RULES OF PERSPECTIVE DRAWING

- Distant objects appear smaller and less distinct than nearer objects.
- Drawings have one (or more) 'vanishing points', where the parallel lines in the drawing appear to converge.
- All objects in the drawing have angular foreshortening and distortion so none of the angles, shapes, or dimensions are 'true'.

For technical applications these rules normally limit the use of perspective views to line drawings only. Once areas of shading and colour are added, it moves away from the category of technical presentation towards that of *art*. The technique of perspective drawing is a means of making a convincing portrayal of a three-dimensional object on a two-dimensional surface. Geometrical perspective drawing is based on sound scientific rules, in contrast to similar techniques in art, which involve varying degrees of interpretation and style. The technique of geometrical perspective makes it possible to construct a visual framework for drawings of building and other architectural features, process plants, and large infrastructure and construction projects. It is, therefore, a well-used (albeit rather complex) technique used in the visual presentation of technical information.

The concept of perspective is based around the observation that linear objects appear to grow smaller as they recede into the distance. Equally, as parallel lines recede, they will appear (because of the laws of optics) to converge to a *vanishing point*. This gives three different types of linear perspective drawings, depending solely on the number of vanishing points, which are termed one-point, two-point, and three-point perspective. They all look different and have different uses.

One-point perspective

A one-point perspective drawing has a single vanishing point so all lines running parallel to the plane of the drawing appear to converge at the same vanishing point on the horizon, in the centre of the drawing. Figure 3.15 shows simple and more complex examples.

Particularly suitable for:

- simple 'landscape views';
- interior architectural views.

A simple building sketch

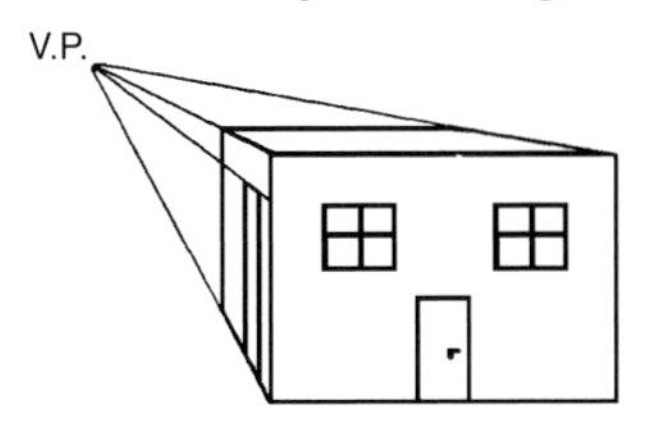

A 'concept sketch' of a product

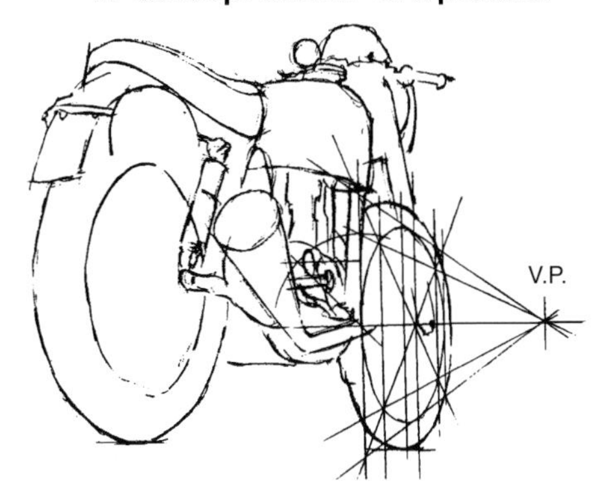

An internal architectural drawing of a building

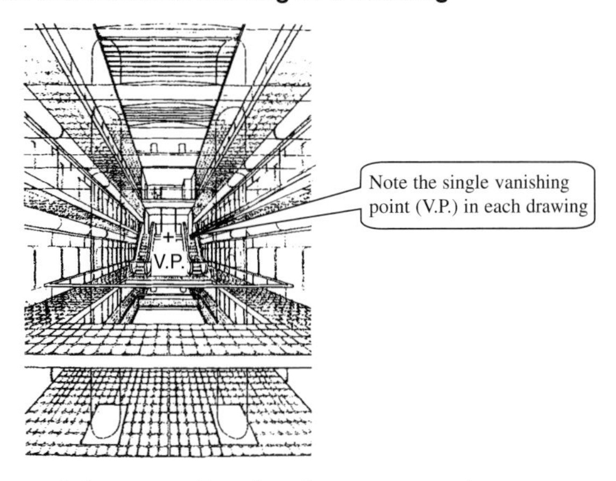

Fig. 3.15 One-point perspective drawings – examples

Two-point perspective

This is a common method for drawings where it is necessary to show detail on two faces of an object. Each side appears to recede symmetrically towards a vanishing point on the horizon on the left and right sides of the object. Two-point perspective is normally only used for objects that have a flat 'ground' plane and vertical uprights. Figure 3.16 shows a typical example.

Particularly suitable for:

- vertical buildings on level ground;
- infrastructure drawings of bridges, dams, etc;
- drawings which contain multiple rows of things: vehicles, houses, assembly lines, etc.

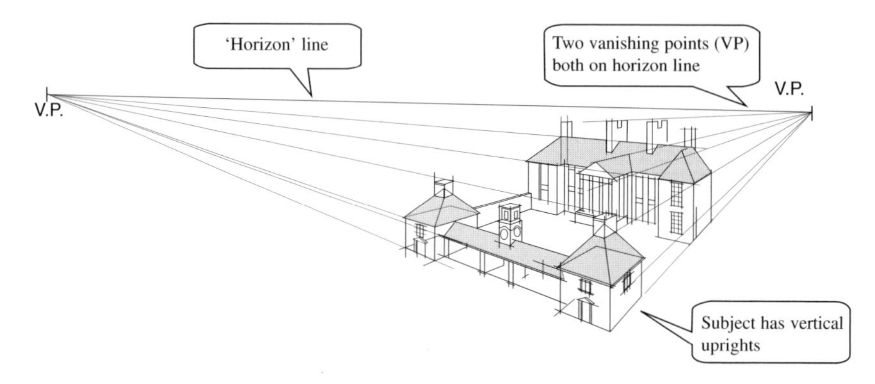

Fig. 3.16 A typical two-point perspective drawing

Three-point perspective

Three-point perspective views are more complex than the one- or two-point types. They are a complex graphical construction containing three principal vanishing points.

> **THREE-POINT PERSPECTIVE VIEWS ARE USED:**
>
> - when the object has none of the planes parallel to the drawing plane;
> - when it is necessary to show *close up views* of large items such as buildings;
> - for views looking steeply *up* at an object (the 'worm's-eye' view) or *down* at it (the 'bird's-eye' view);
> - for dramatic visual effect, emphasizing shape and scale.

Figure 3.17 illustrates the main use of three-point perspective views, i.e. for showing architectural features such as tall buildings.

The 'worm's-eye' three-point view

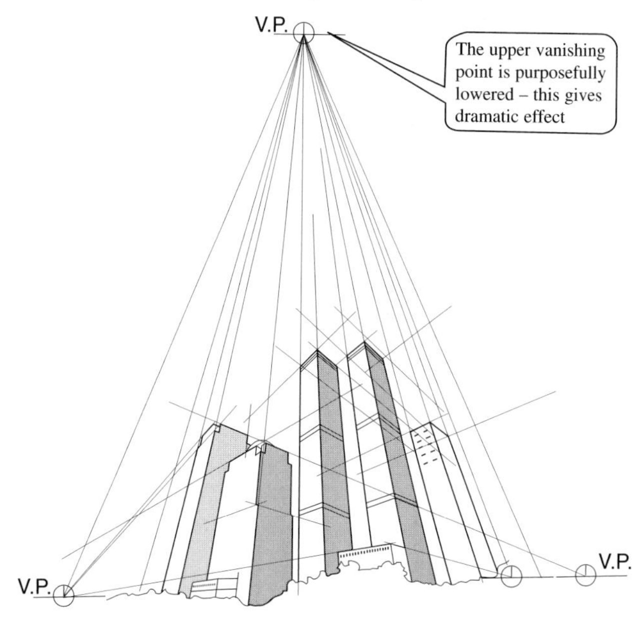

The 'bird's-eye' three-point view

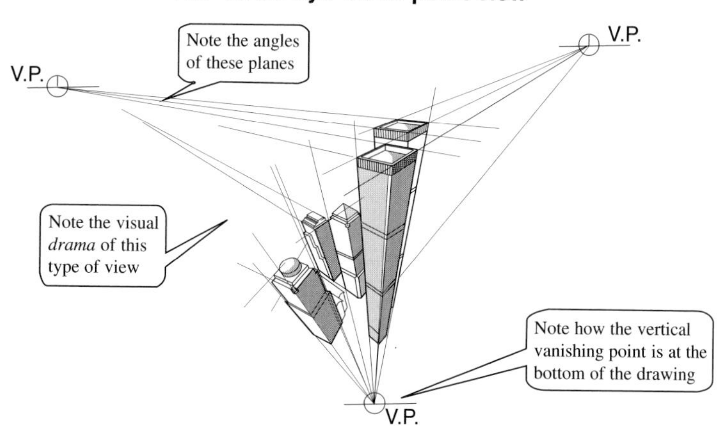

Fig. 3.17 Three-point perspective views of buildings

Useful references

http://www.cinderella.de/demo/gallery/Perspektive.html
http://forum.swarthmore.edu/sum95/math_and/perspective/perspect.html

Axonometric projection

Axonometric projection is a specialist way to show 3-D objects. It is also used for aerial views, mainly of buildings, when is it effectively a combination of two different separate aerial view techniques:

- *Vertical perspective views* – looking vertically down perpendicular to the earth's surface. Such images are common in mapping and survey disciplines and they can be overlapped to provide a geographically accurate representation of a large area.
- *Oblique perspective views* – looking at objects on the earth, e.g. buildings, from their sides. This provides better three-dimensional information than the vertical perspective view and looks more natural. The disadvantage is that objects in the distance appear smaller and, therefore, lack detail.

An axonometric view combines the advantages of the vertical and oblique perspective views. It does this by combining a geographically accurate vertical image of the area with a constant perspective side view of the 3-D objects on the earth's surface. Figure 3.18 shows two examples.

A (simplified) axonometric view – room and furniture

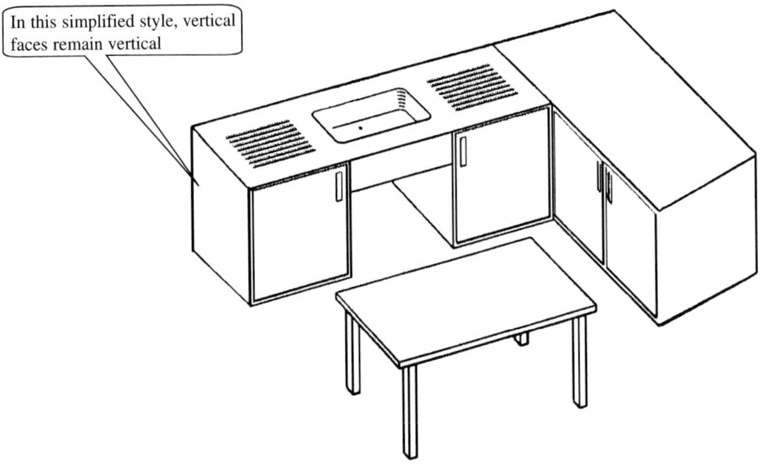

Fig. 3.18(a) Examples of axonometric views

An axonometric 'cityscape' view

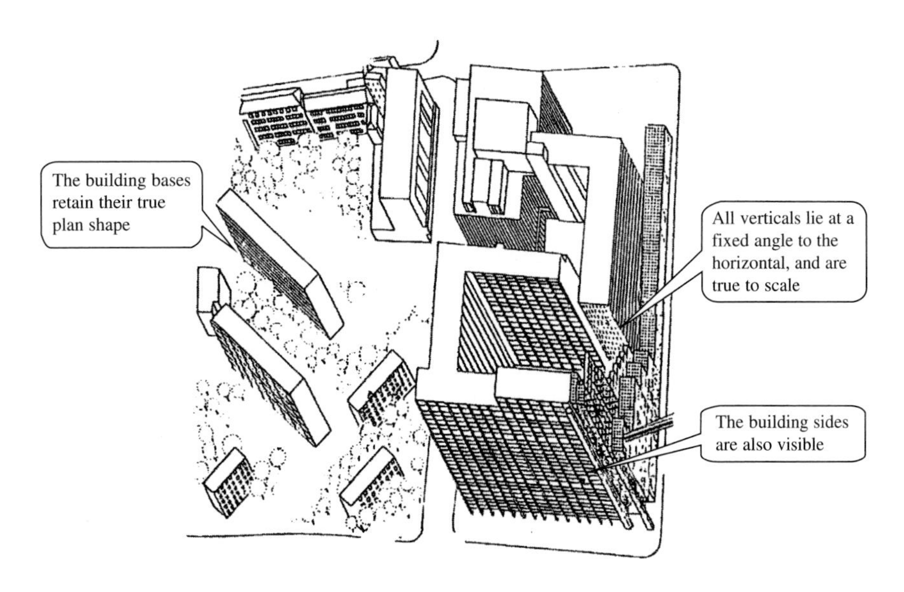

Fig. 3.18(b) Examples of axonometric views

Useful references

http://www.architecture.ubc.ca/eds95/group3/axon2.gif
http://www.galaxymaps.com
http://www.astro.virginia.edu/~eww6n/math/ProjectiveGeometry.html

Engineering drawings – their function

We saw at the beginning of this chapter that technical drawings belong to part of a family or set. Each member of the set has a particular purpose, but only when they are taken together do the drawings provide a unique description of all the technical details of an object. The individual types of drawings that make up the set follow the pattern shown in Fig. 3.19. This arrangement is by no means exhaustive, but is a good general description of the drawing types that you will find in use in technical/engineering disciplines.

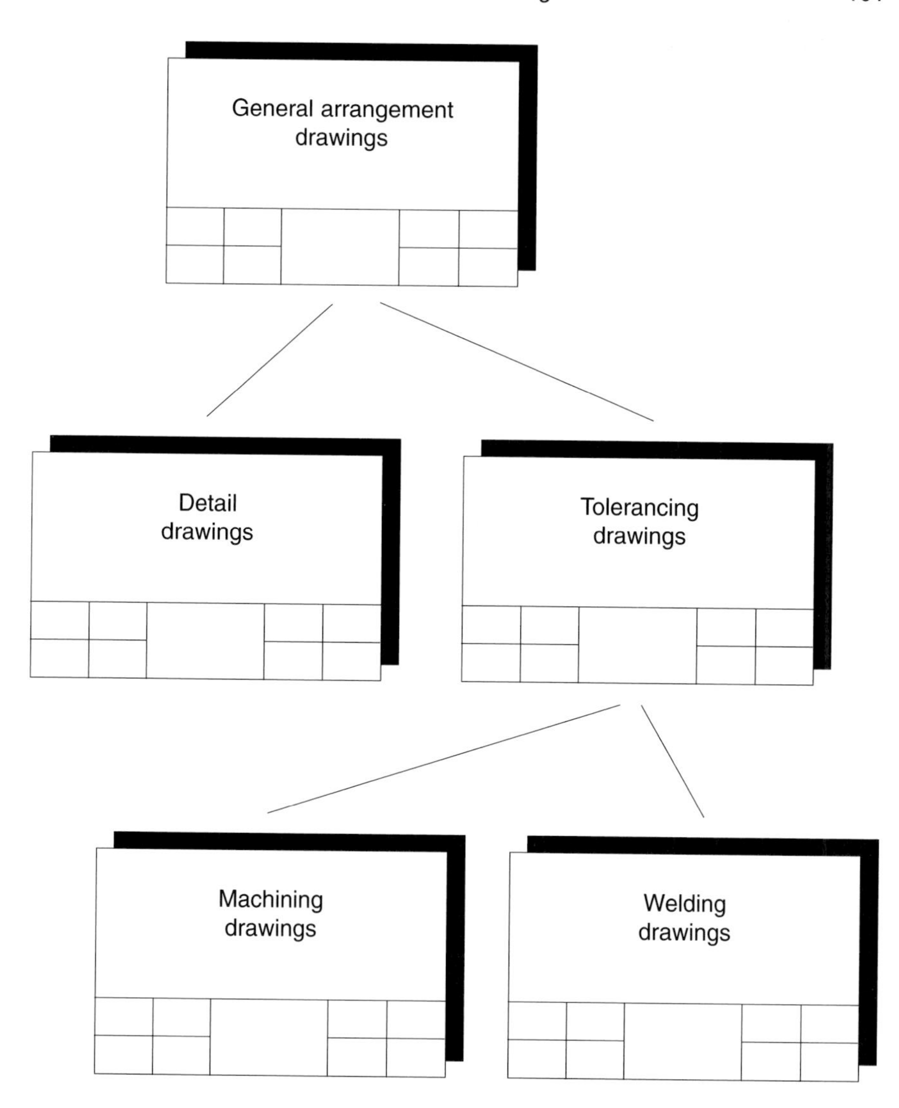

Fig. 3.19 A 'set' of individual engineering drawings

General arrangement (GA) drawings

GA drawings show a general view of an engineering object. The features are:

- the object is shown assembled (if it is made from several components);
- each drawing contains a material list, showing what the components are made of;
- some overall dimensions may be shown but not enough to enable the object to be manufactured;
- the drawing references a list of other, more detailed, drawings of the component.

The GA drawing is widely accepted as being the first drawing to look at if you want to 'get the general picture' of a technical object or engineering component. Figure 3.20 shows an example for a ratchet and pawl assembly.

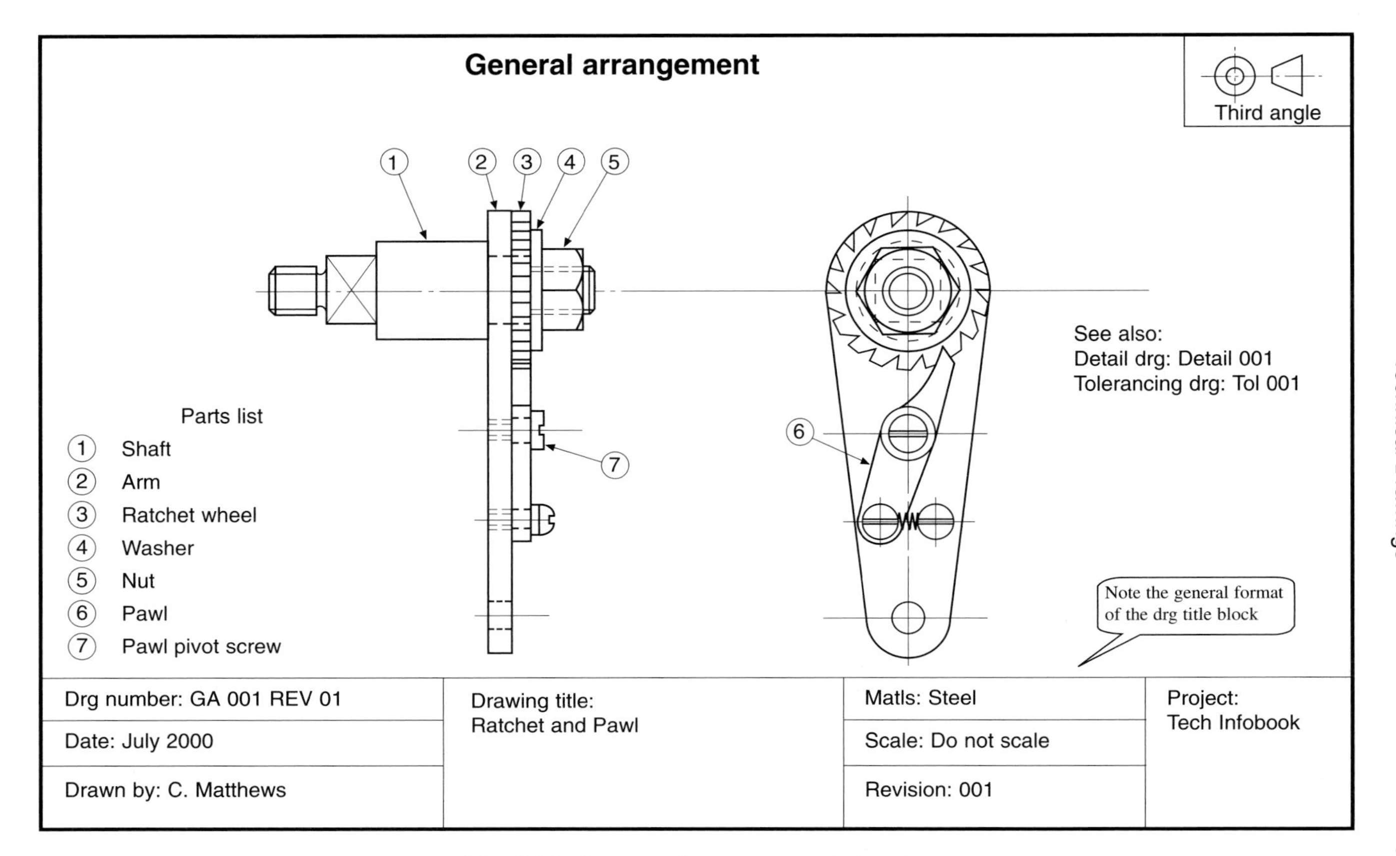

Fig. 3.20 A general arrangement drawing

Detail drawings

These contain full information about an object; sufficient to manufacture it if required. Their features are:

- they only contain information about the object(s) in question – no general information about other, related components;
- they are dimensioned;
- they often contain manufacturing information.

Figure 3.21 shows a typical example of detail drawings, in this case for three of the components of the ratchet–pawl assembly shown in Fig. 3.20.

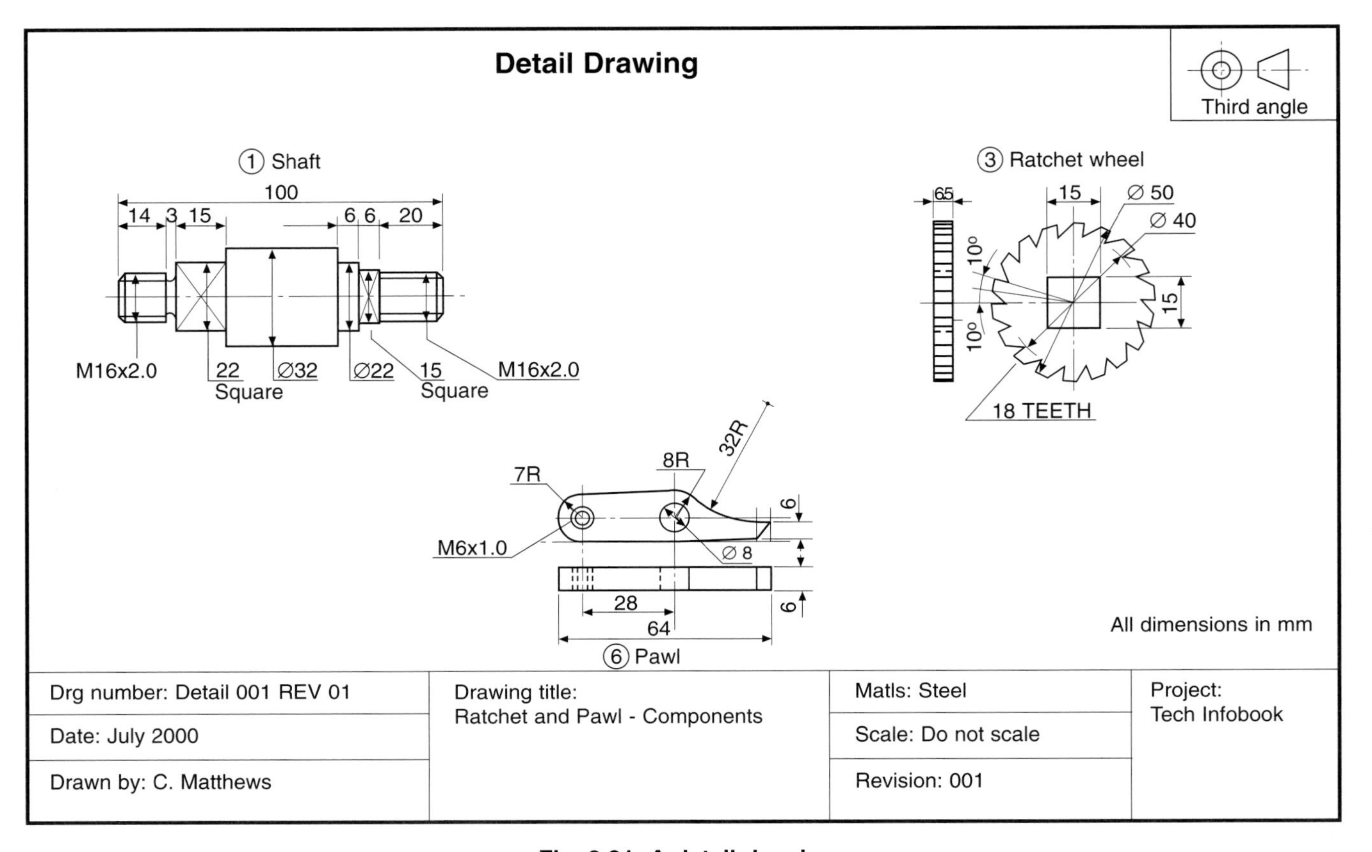

Fig. 3.21 A detail drawing

Tolerancing drawings

Manufacturing and assembly tolerances are an important part of any engineering component, particularly those with moving parts. Tolerances are traditionally shown on the detail drawings or, for very complex items, on separate tolerancing drawings. The conventions for tolerancing are well developed, and are referenced in published technical standards.

Figure 3.22 shows typical tolerancing information for one of the rachet–pawl mechanism components, in this case based on the conventions and practices of BS 4500 **(2)** and BS 308.

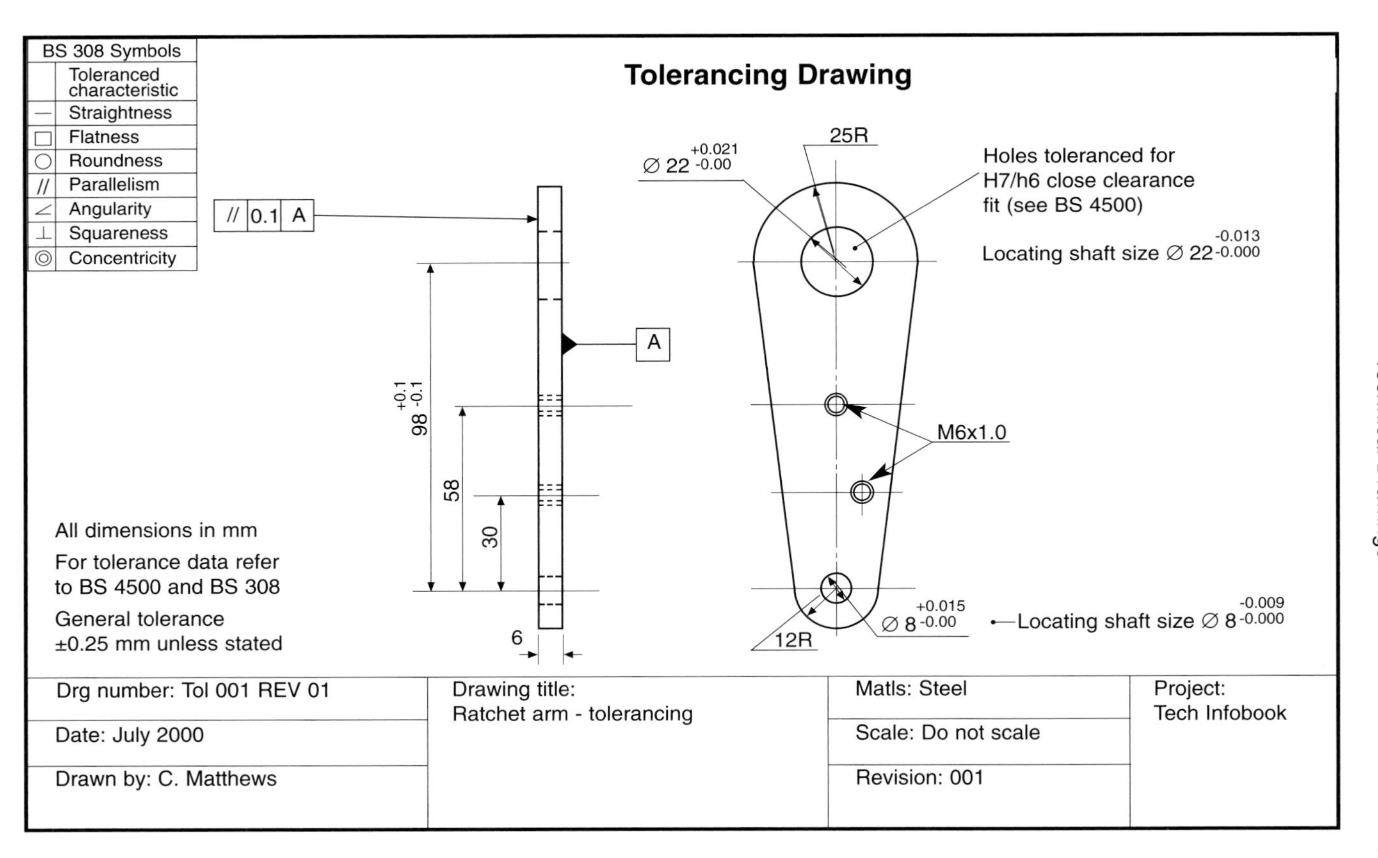

Fig. 3.22 A tolerancing drawing

Machining drawings

These are a subset of the detail drawings of engineering components. Their purpose is to carry critical instructions about how to machine a component when it is being manufactured. Machining drawings contain technical information about:

- the type of machined surface;
- surface finish and texture (roughness and smoothness);
- the 'lay' of a machined surface (the pattern of tool-marks left after machining);
- machining tolerances.

Figure 3.23 shows a simplified example of a machining drawing of a precision engineering component.

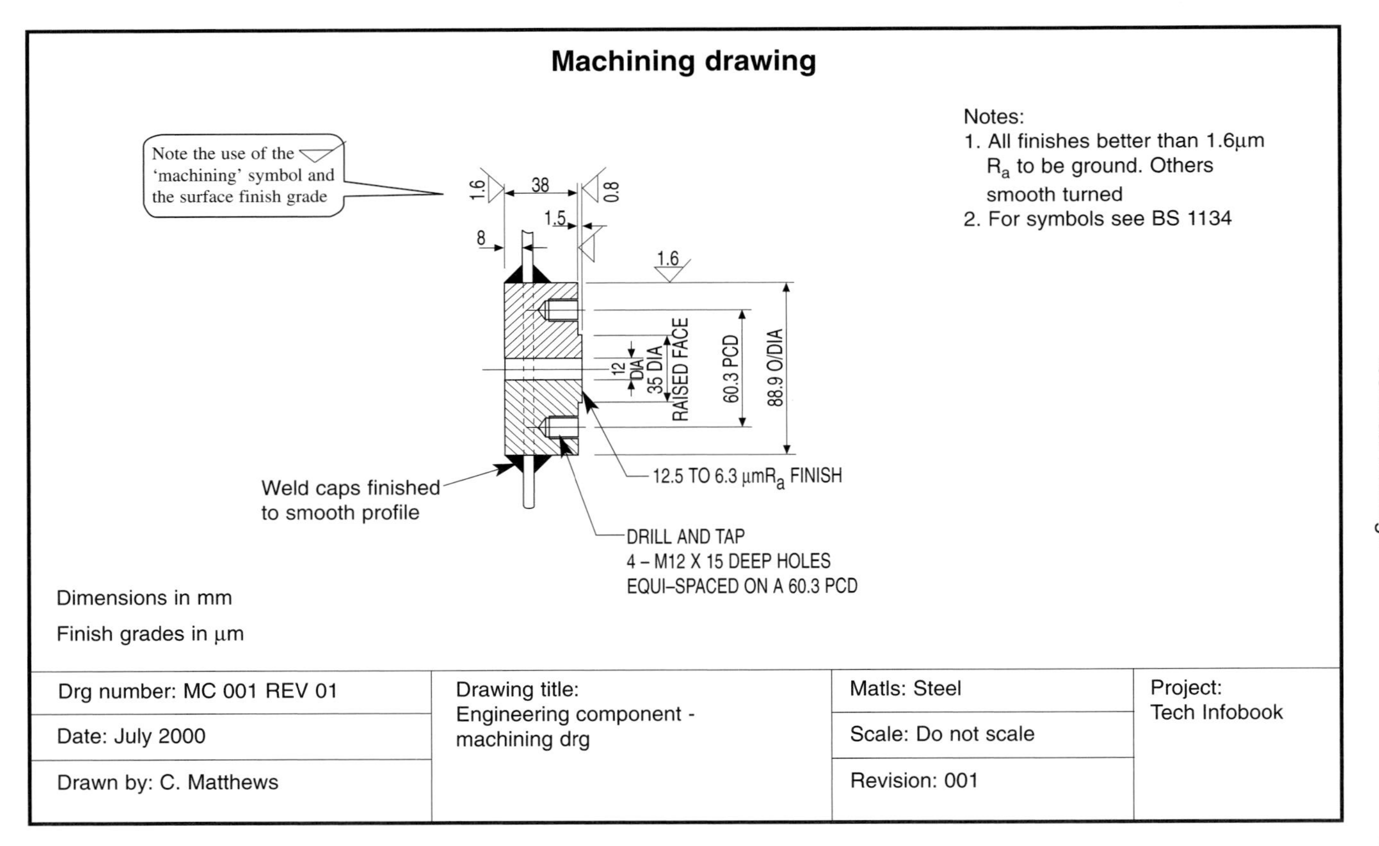

Fig. 3.23 A (simplified) machining drawing

Welding drawings

These are similar to detail drawings and in common use for all fabricated components and structures. Their purpose is to convey instructions about how a structure is welded together during manufacture. Expect to see:

- layout and orientation of weld preparations and the welds themselves;
- cross-references to other documents such as weld procedure specifications (WPSs) and weld procedure qualification records (WPQRs);
- standard symbols from published technical standards (see BS 499 (**3**)).

Figure 3.24 shows a simplified example of a welding drawing for a fabricated stainless steel bobbin.

Useful references

http://www.hlbtech.com/weld.htm

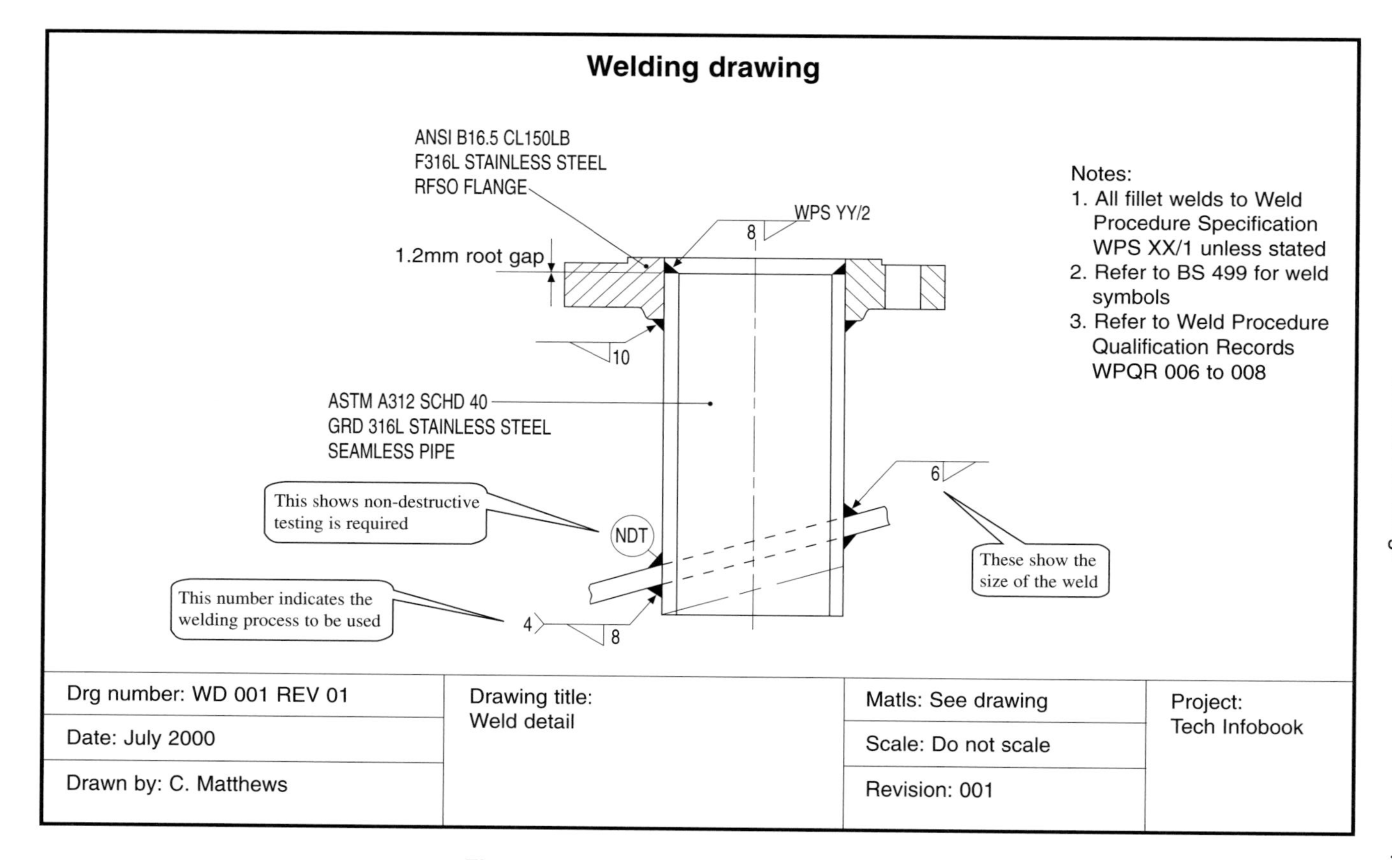

Fig. 3.24 A (simplified) welding drawing

And then: fanciful geometry – drawing the tesseract

It is nothing but convention that the most complex technical drawings we can manage are presented using three-dimensional geometry. There is no scientific proof that space is three-dimensional, because the pure mathematics we use is only bothered about its own logical consistency. Hence space can only be considered as being 'n' dimensional – a pretty useless conclusion.

Maybe there are four dimensions. The notion of a fourth dimension is abstract and lives in the purest realm of conception, as do many other well-accepted mathematical notions. When was the last time you saw the square root of minus one apple, for example?

Four-dimensional (4-D) geometry can be thought of as a system of drawing which uses four sets of *co-ordinates*. The nearest that anyone has come to drawing a 4-D object is shown in Fig 3.25, known as a hypercube or *tesseract*. A tesseract is a four-dimensional analogue of a three-dimensional cube. It is not, of course, a true representation because in the same way that a three-dimensional object can only be drawn as a perspective view on two-dimensional paper, a two-dimensional picture of a four-dimensional object is a *perspective of a perspective* – a poor representation at best. Not only is it difficult to draw, it is also hard to visualize.

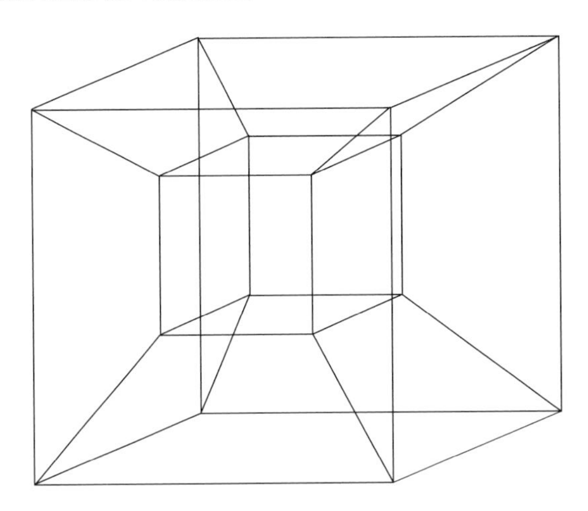

The Tesseract is 'bounded' by 8 cubes and has:

- 16 vertices
- 24 faces
- 32 edges

Fig. 3.25 Fanciful geometry – the tesseract

Difficult concept though the tesseract is, it is a gateway to the elusive fourth dimension. Can you draw a better one?

Useful reference

http://www.inconcept.com/JCM/December1998/index.html

References

(1) BS 308: 1993 *Engineering Drawing Practice* various parts. This is a similar standard to ISO 216:1973.

(2) BS 4500: 1985 ISO *Limits and Fits* various parts.

(3) BS 499: 1992 *Welding Terms and Symbols.*

Chapter 4

Conceptual Technical Design

Introduction – conceptual design

The process of technical design involves many challenges for the presentation of technical information. The process is a complex one and there are many different, often conflicting, views on how it is done and the steps it involves. The early stages of any design activity, however, are conceptual and involve the creative thought processes that provide the innovative 'sparks' which will eventually translate into the physical design features of the finished object. One popular view is that these early stages, and the ones that subsequently follow it, are part of a systematic activity. *Systematic design* conceives the design process as a series of linear steps contained within a total context or framework. It is a well-ordered and structured approach to the design process.

Presenting systematic design information

The process of systematic design involves a set of techniques for presenting technical information at the various stages of the process. Some of these techniques are mechanistic, while others act more as aids to creativity, serving to tie down half-formed concepts and technical ideas. This chapter shows the main techniques that are in use, and presents them in broadly chronological order. Remember that, because of the conceptual nature of the subject at this stage, these presentation methods are not rigid – there is room for variations and adaptations.

Synthesis sketches or 'notemaps'

These are a way of representing ideas and their interrelationships in graphical form. They are an attempt to understand and portray real-world technical complexity on a sheet of paper, in a way that allows this understanding to be conveyed to other people. Figure 4.1 shows an example used in the process of *design synthesis* – in this case for simple robotics. Remember that the diagram is not sequential or chronological; it is a *conceptual* network.

Design considerations in robotics

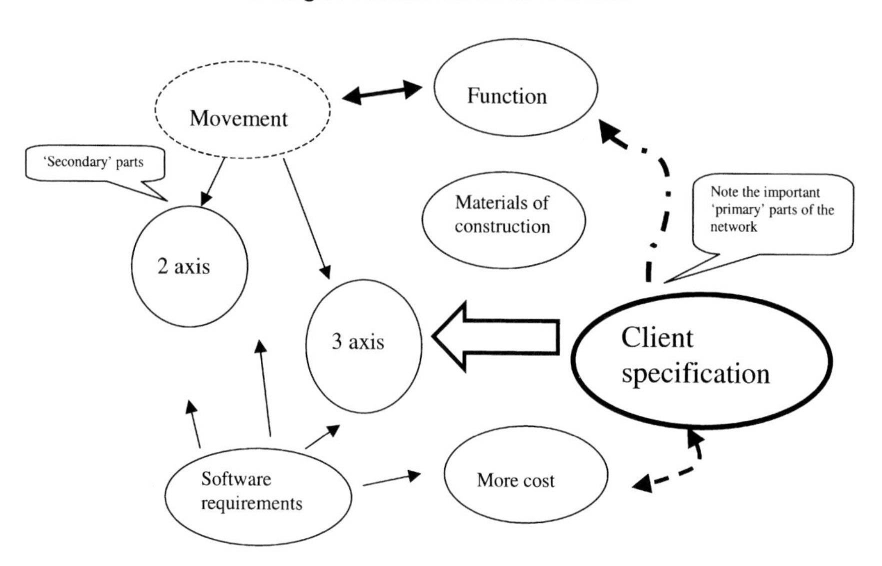

Fig. 4.1 A synthesis sketch or 'notemap'

NOTEMAPS – TECHNICAL TIPS

- Note how the network has primary and secondary parts; these are added as the ideas develop.
- Use different thicknesses of line. Thick lines represent clearer, more direct links between concepts and ideas.

Task clarification diagrams

These are a formal method of clarifying tasks, used after the synthesis sketch (notemap) stage of the systematic design process. They help to formalize exactly what the tasks of the design process are and so help with early division of tasks and the allocation of responsibilities to people in a design team. The diagrams visualize the tasks as a series of vertical 'stairs' (see Fig. 4.2).

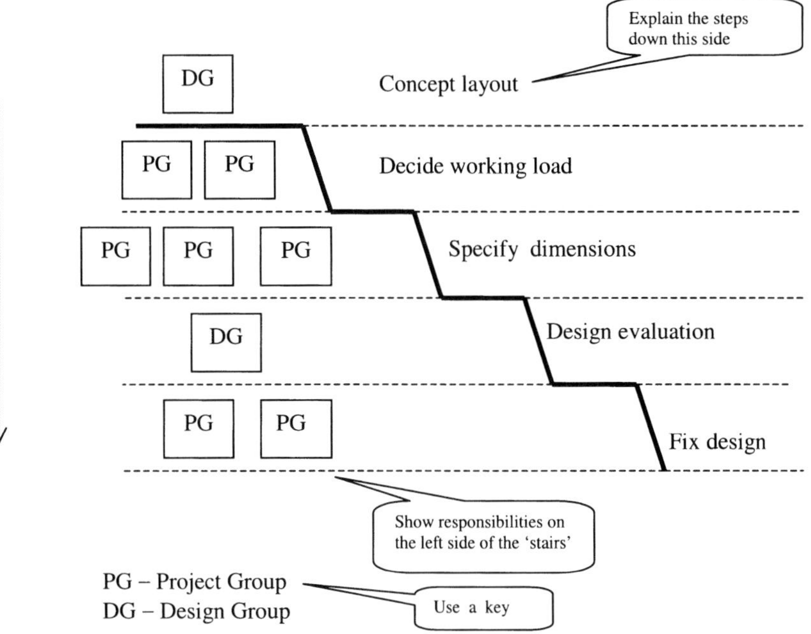

Fig. 4.2 A task clarification diagram

> **TASK CLARIFICATION DIAGRAMS – TECHNICAL TIPS**
>
> - Note the inferred time axis in the vertical direction (but do not show it).
> - Provide an explanation of the stages on the right-hand side of the 'stairs', matching up with the 'boxes of responsibility' on the left side.
> - Do not make this diagram too complicated; it is for design task *clarification only*, not to explain in detail what the tasks are.

Selection algorithms

Similar in style to task clarification diagrams, these are a way of showing the breakdown of the design process into its component parts or steps, see Fig. 4.3. The steps are expressed in algorithm form with the content of the algorithm 'boxes' chosen to suit the nature of the design process. Note how the algorithm chain is subdivided into several phases for clarity (and to fit in with any previous decision on task clarification).

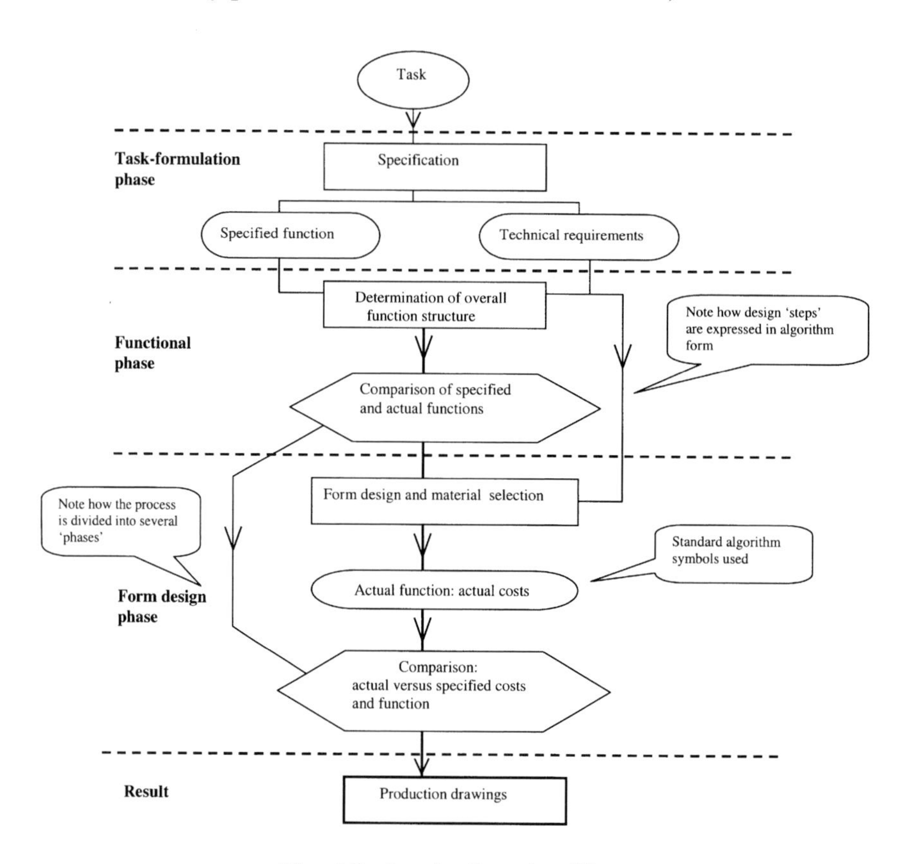

Fig. 4.3 A selection algorithm

SELECTION ALGORITHMS – TECHNICAL TIPS

- Use accepted algorithm symbols such as those in Fig. 4.3 to differentiate between statements of function, comparison activities, etc.
- Do not forget arrowheads, showing the path around the algorithm diagram.
- Work from the *task* at the top of the diagram to the *product* (or design) at the bottom, not the other way round.

System representations

It is common in the conceptual stages of the design process to represent a design or product as a *linked system of functions*. This idea of a system is a useful way to start the design of many types of process plants, machines, and equipment – in fact functional products of almost any type. Figure 4.4 shows a typical example for a laptop computer. Note how the functions of the item are expressed as a number of systems and subsystems within the main system boundary. The separate parts are then joined by links, showing how they interrelate with each other.

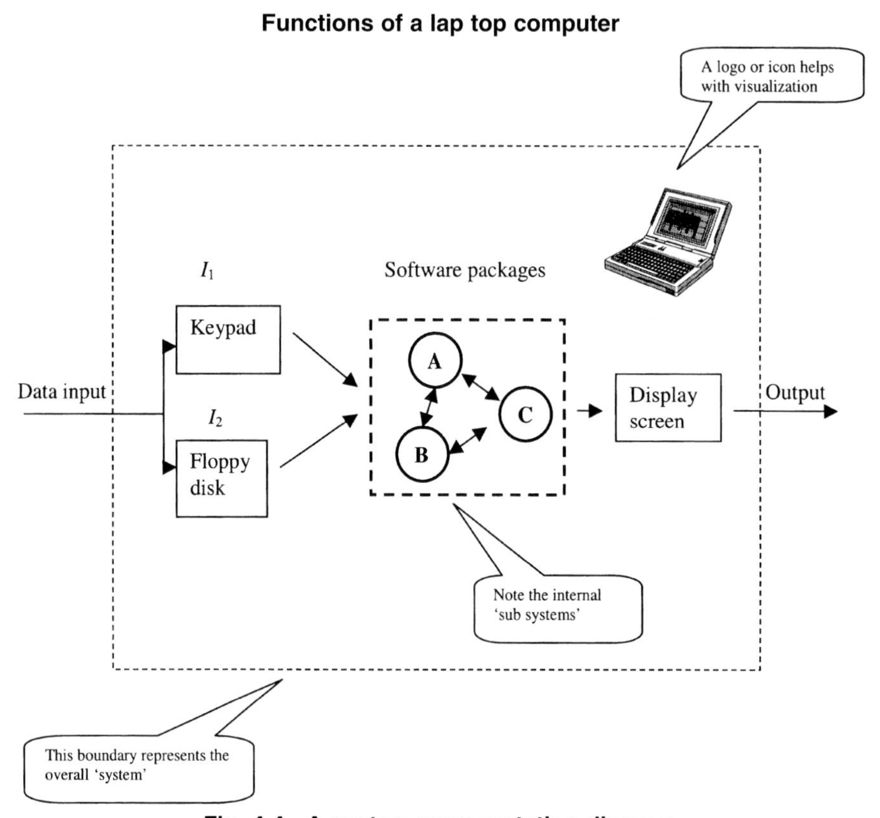

Fig. 4.4 A system representation diagram

SYSTEM REPRESENTATIONS – TECHNICAL TIPS

- If it is to make sense, a system diagram must have inputs and outputs.
- Note how the systems are given letter or number designations. This is easier than trying to find accurate names during this conceptual stage of the design process.

Useful reference

http://www.vensim.com/venplus.html

http.nyquist.ee.ualberta.ca/~wjoerg/SE/SE_Notes-F/SE_What-F/SE_Req-F/?SE_SysMod-F/SE

Function structure diagrams

Function structure diagrams follow on from system representations. Their purpose is to help designers think clearly about the number and type of sub-functions that are needed to fulfil the objectives of a particular design. They are also useful in identifying design variants (slightly different ways of doing the same thing). These diagrams are in common use in the design of almost any manufacturing/production-related process such as batch manufacturing and mass production, and a wide range of project work involving human activity.

The diagrams come in a set of two. Step 1 shows the structure of functions and sub-functions in 3-D form, mainly for the purpose of visualization (Fig. 4.5(a)).

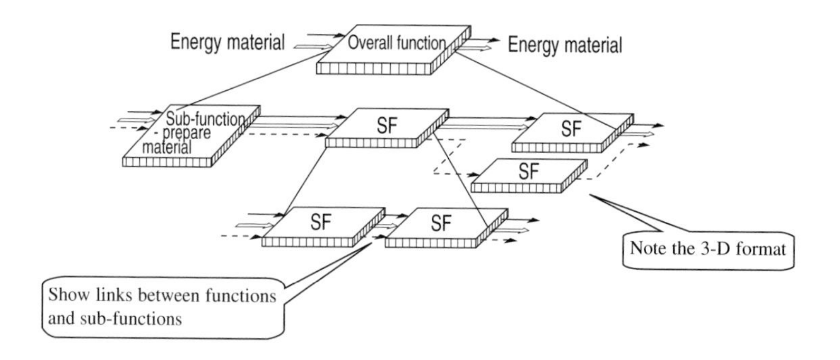

Fig. 4.5(a) The function structure diagram – step 1

Step 2 (Fig. 4.5(b)) displays the technical activities in each sub-function in indivisible blocks, i.e. it shows which activities cannot be subdivided, because of the physical nature of the activities themselves.

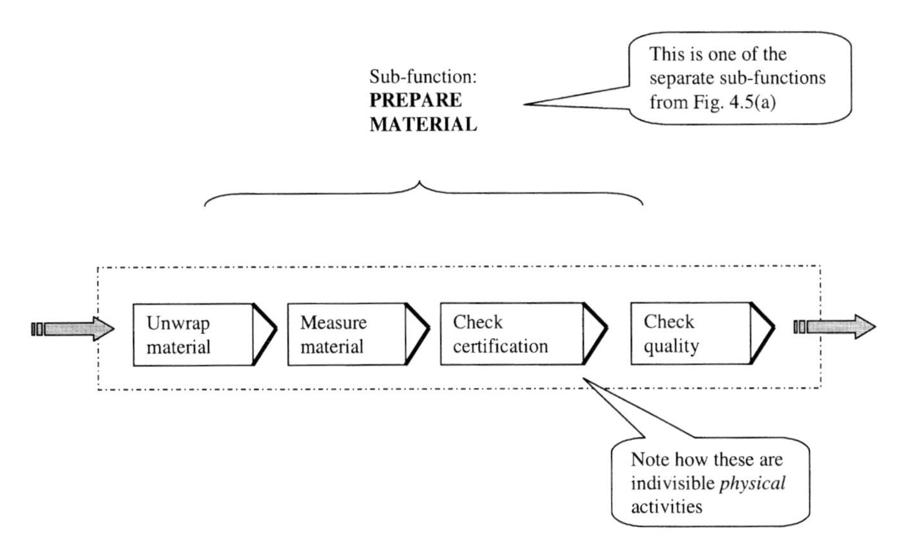

Fig. 4.5(b) The function structure diagram – step 2

Design-solution matrices

Design-solution matrices combine tabular and graphical principles. Their purpose is to show physical *principles* of solutions to design problems, rather than the detail of their solutions (that comes later).

Main uses are in:

- mechanical engineering;
- electrical engineering;
- process engineering;
- control engineering.

Each design sub-function (left-hand column of Fig. 4.6) is analysed in the subsequent columns and an outline solution provided in the form of an equation or sketch. Sketches are important because they help with visualization of the physical solution to a design problem. These matrices can be very large, covering all the principles of a complex technical item.

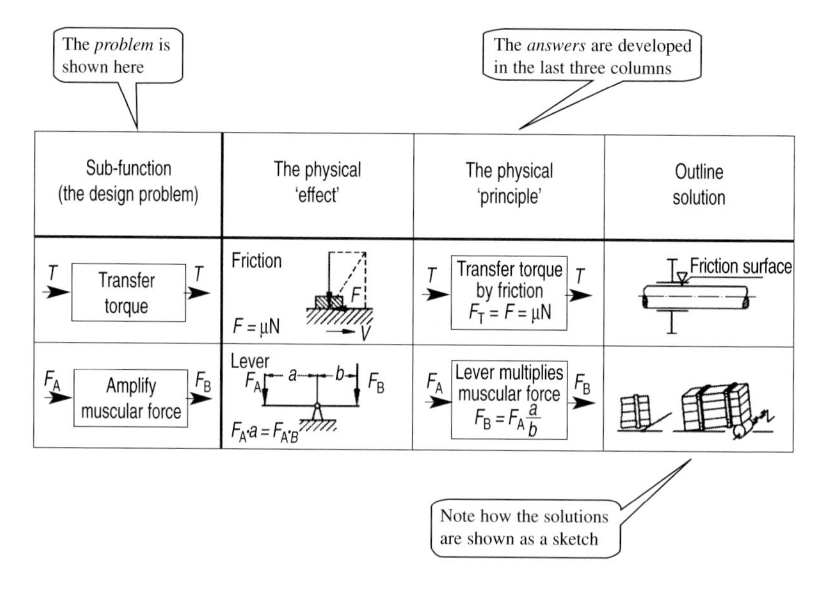

Fig. 4.6 A design-solution matrix

Useful reference

http://www.wsdot.wa.gov/eesc/design/dsgnmatrix.htm

Form interrelationship charts

Form interrelationship charts are the follow-on in the design process after solution principles have been decided. They allow practical consideration and comparison of the physical form of the solution principles but in a way that continues to let ideas develop. Figure 4.7 shows the idea for the design of a shaft/sleeve fixing. Note how the presentation is not hierarchical or systematic in any way, but rather it lets things develop in a series of forward design steps until all options have been considered.

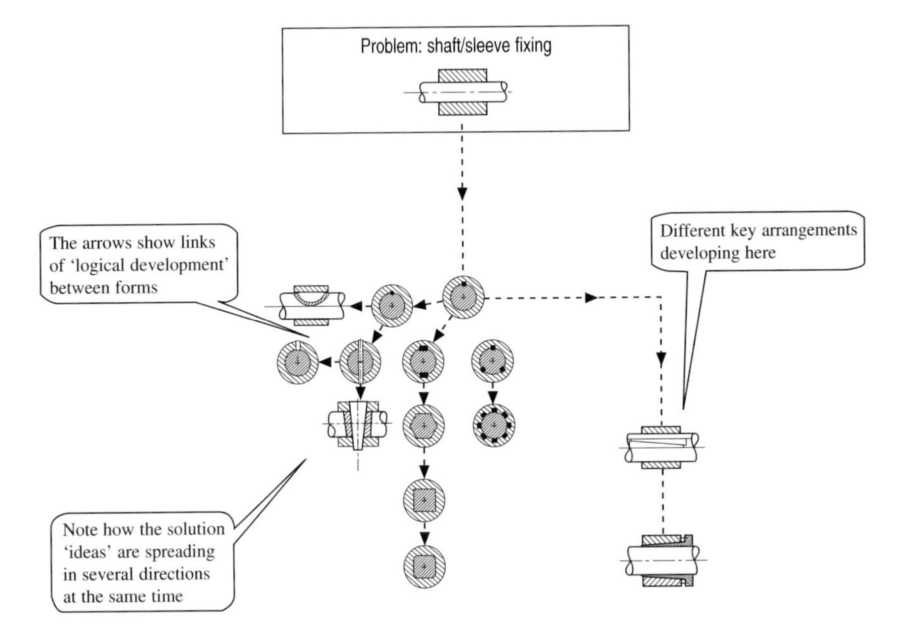

Fig. 4.7 A form interrelationship chart

FORM INTERRELATIONSHIP CHARTS – TECHNICAL TIPS

- Use arrows to show how the ideas have developed. This will be useful when you have to explain the chart to other people.
- Do not make the sketches too complicated. Aim for something like those in Fig. 4.7 which are sufficient to show the form of the various design solutions.

Systematic selection charts

These are an advanced form of matrix analysis. The technique allows a quantified assessment of multiple design features, enabling designers to eliminate some and express preferences for others. There can be large amounts of information included in systematic selection charts (Fig. 4.8) This means that they warrant detailed analysis. They are an exhaustive, systematic method of weighting and selecting design features. They are not a quick-fix assessment method.

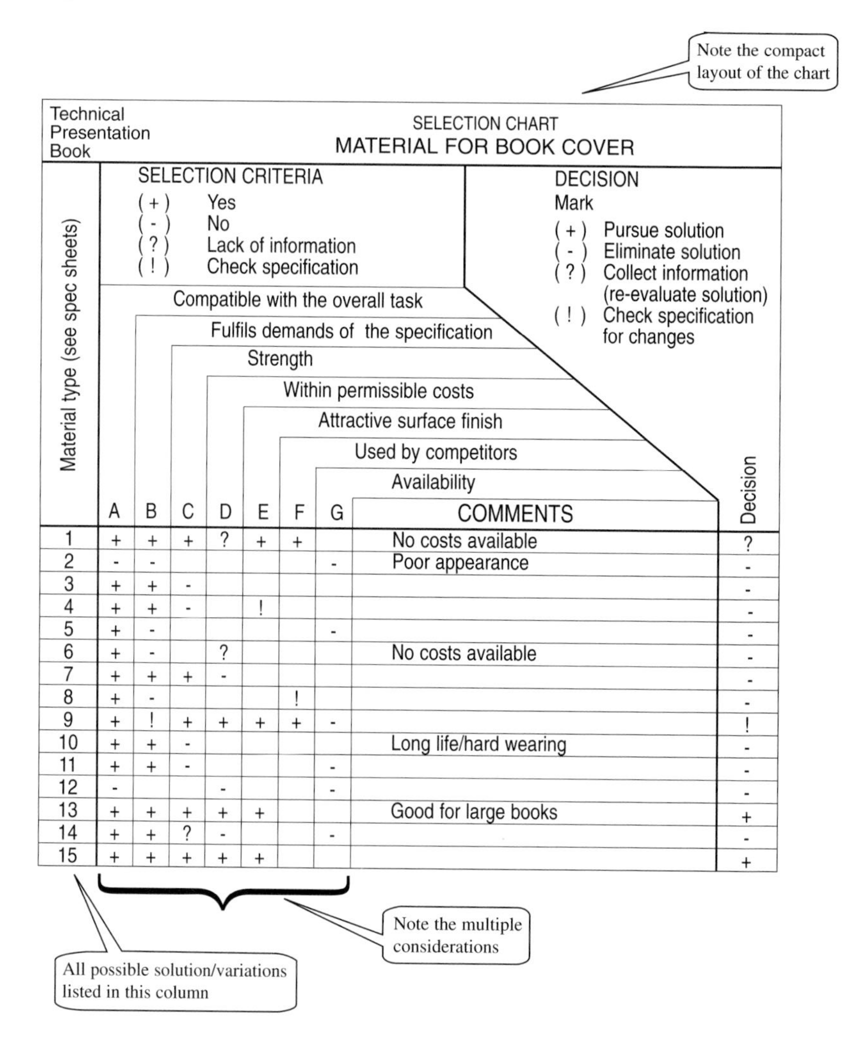

Material type	A	B	C	D	E	F	G	COMMENTS	Decision
1	+	+	+	?	+	+		No costs available	?
2	-	-					-	Poor appearance	-
3	+	+	-						-
4	+	+	-		!				-
5	+	-					-		-
6	+	-		?				No costs available	-
7	+	+	+	-					-
8	+	-				!			-
9	+	!	+	+	+	+	-		!
10	+	+	-					Long life/hard wearing	-
11	+	+	-				-		-
12	-			-			-		-
13	+	+	+	+	+			Good for large books	+
14	+	+	?	-			-		-
15	+	+	+	+	+				+

SYSTEMATIC SELECTION CHARTS – TECHNICAL TIPS

- The layout in Fig. 4.8 has been developed to show the maximum possible selection information in a given space. Use it.
- Note the connections with solution variants charts shown earlier (Fig. 4.6). Remember that systematic selection charts are *part of* the process of conceptual design, not a stand-alone technique.

Component selection diagrams

Although partially practical, these diagrams are sometimes used during the conceptual design stage to help focus designers' minds on the later selection of components. Their purpose is to show the issues that influence the selection of an engineering component. This may be useful for later justification purposes, or when it is necessary to look again at why certain choices were made. The diagram is shown in the form of a simple linear flow chart containing statements and 'decision' questions (Fig. 4.9). There may be one or more conclusions, depending on the context.

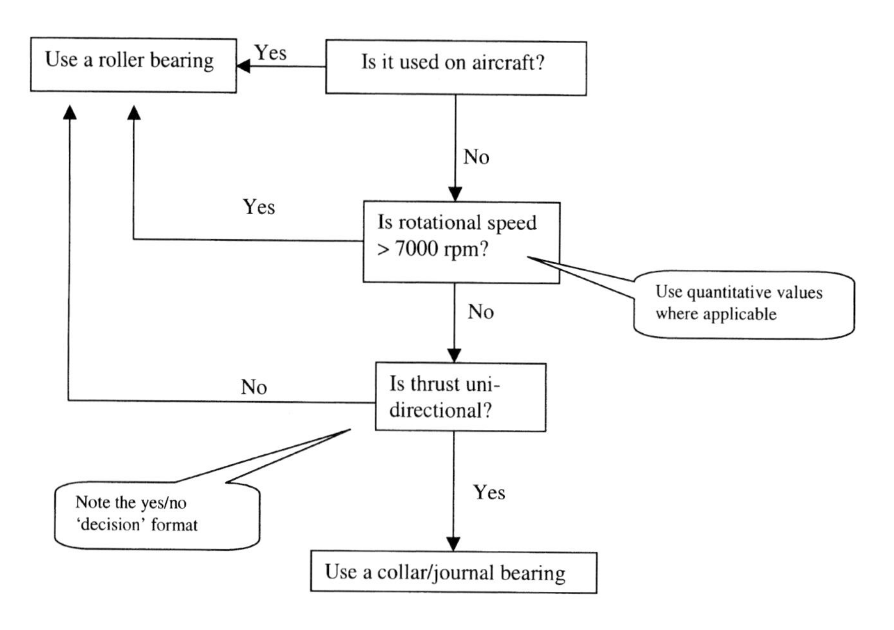

Fig. 4.9 A simplified component selection diagram

SIMPLIFIED COMPONENT SELECTION DIAGRAMS – TECHNICAL TIPS

- Note the yes/no format of the flowchart; this is the conventional way to show decisions in this type of diagram.
- Note the use of quantitative values (e.g. > 7000 rpm?). This type of design accuracy is important when considering physical engineering components. Do not let these diagrams stay *too* conceptual.

Chapter 5

Practical Technical Design

Introduction – practical design

The thing that differentiates practical design from purely conceptual design is its degree of tangibility. Whereas conceptual ideas have the option of being presented in written form, or merely kept in the designer's head, practical design always involves expressing technical ideas on paper or computer to other people. Hence the way that this technical information is presented is important; presentation is a fundamental part of the practical design process.

Are there any similarities with the conceptual design process?
Yes. All the stages are different and each offers a choice of several ways to do the same thing. The stages are, as with conceptual design, not always exactly sequential even though they follow a basically chronological pattern. In contrast though, practical technical design uses presentation techniques that are more conventional than those of conceptual design; you will see that they have a more practical engineering 'feel' to them.

TIPS ON USING THIS CHAPTER

- Remember the *multidisciplinary* application of the presentation methods shown. They have many uses, not just in the technical disciplines used in the illustrations shown in this book.
- All presentation techniques will not be applicable to all disciplines – it depends on their suitability to the subject, and the context.
- And the key point – do not forget to use imagination. None of the techniques are so rigid that you cannot adapt them to suit yourself.

The portfolio of design sketches

Most components and products start their life as design sketches. There is a basic set of four types of sketches used in the early design stages, each with its own specific purpose. These are:

- initial ideas sketches;
- initial detailing sketches;
- ergonomic/anthropometric sketches;
- final detailing sketches.

Initial ideas and detailing sketches

These encapsulate the early ideas about what the product needs to do (its function) and broadly what it looks like. An outline technical specification is also sometimes included on the drawing. Figures 5.1 (a) and (b) show typical examples.

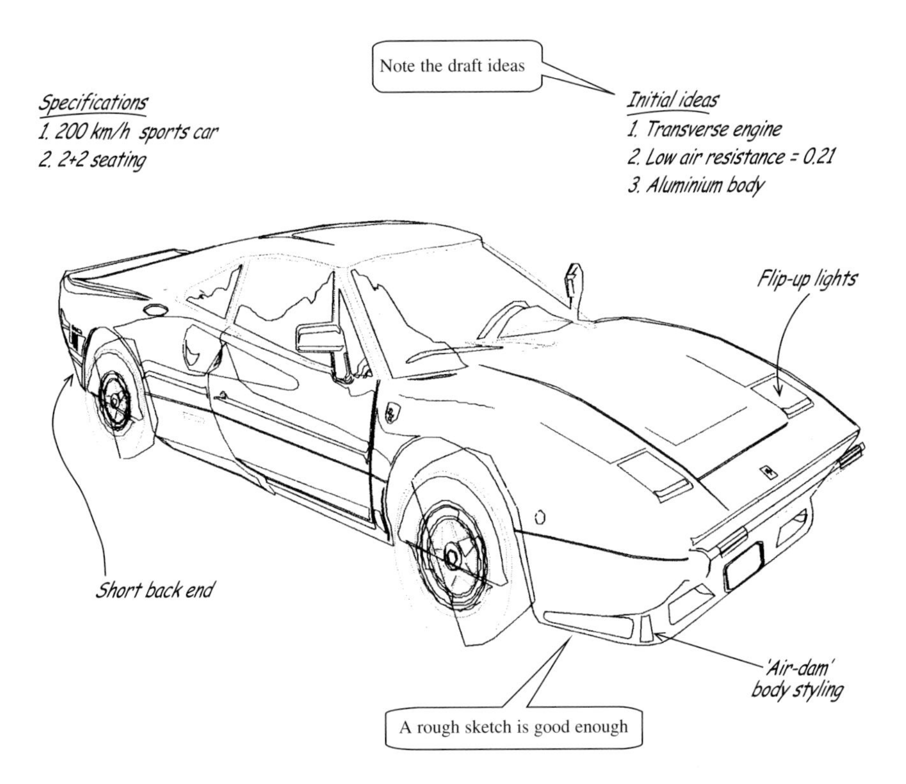

Fig. 5.1(a) The initial ideas sketch

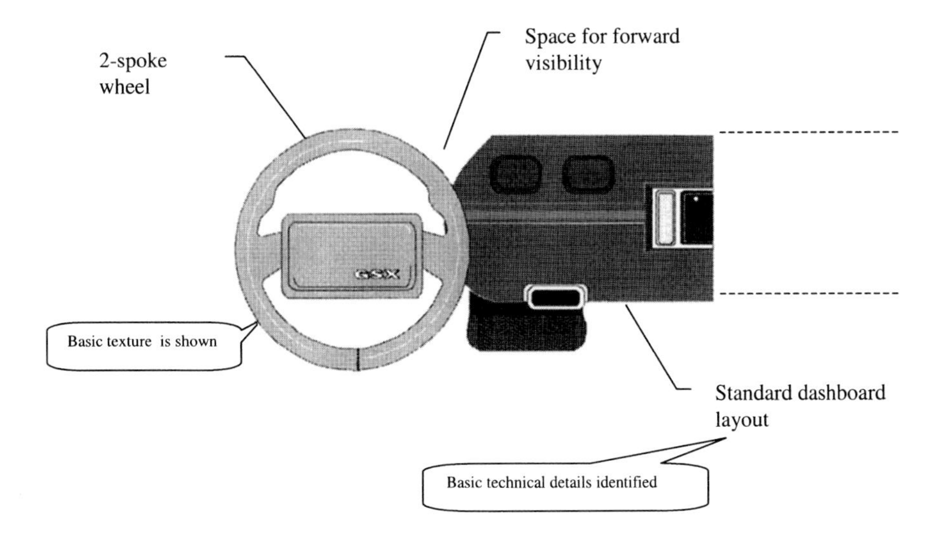

Fig. 5.1(b) The initial detailing sketch

Ergonomic and anthropometric sketches

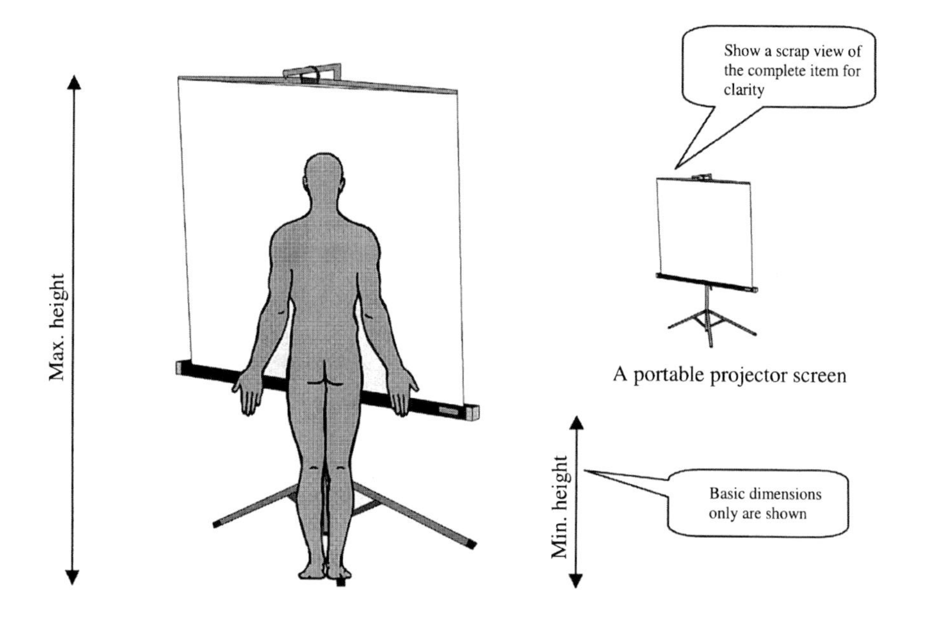

Fig. 5.2 A basic ergonomic sketch

Ergonomic sketches are about 'human factors engineering' of the product design. They show key dimensions and features of the design related to the way that it interfaces with humans in areas such as sight, hearing, attention span, mental and physical workload, skill, and fatigue. Anthropometric sketches are a sub-set of the ergonomic sketches (in reality ergonomics and anthropometry are overlapping subjects) which cover the 'fit-to-body' of the design. Factors such as portability (can someone lift it?) are presented in this form. Figure 5.2 shows an example.

Final detailing sketches

These show final detail of important design features such as joints, pivots, fixings, etc. Their purpose is to provide sufficient information to allow a detailed orthographic 'engineering' drawing of the key areas, so that features such as tolerances, clearances, and surface finish can be specified, see Fig. 5.3.

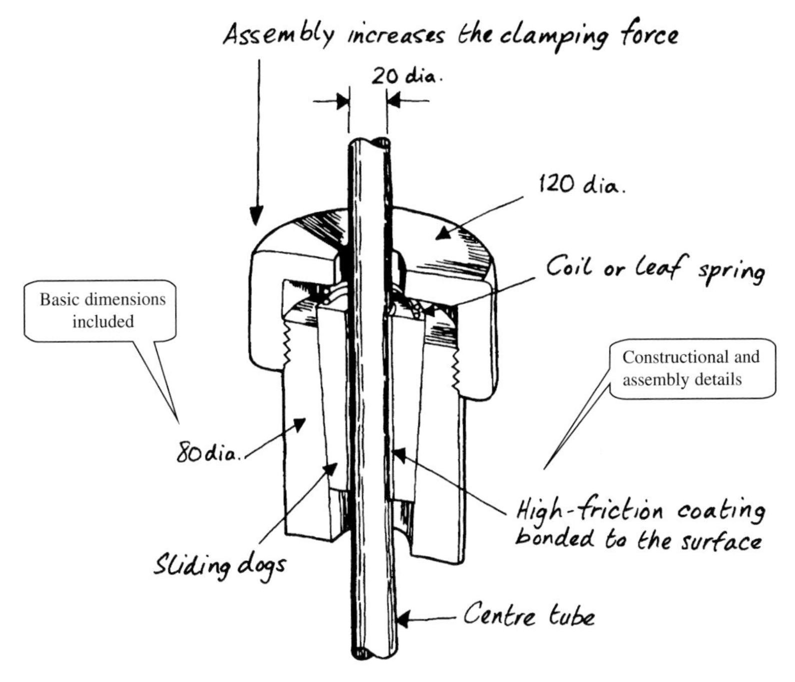

A friction-operated tube-pulling dog

The product analysis drawing

A product analysis drawing is an amalgamated and simplified version of the set of design sketches shown previously. It is simple enough to interest the casual viewer, as well as experienced technical people, who may be involved in the design process. Note the key features (Fig. 5.4):

- simple illustrations without formal technical detail;
- lots of explanations as text on the drawing;
- general discussions and comparisons to make the drawing a separate little story.

This type of drawing is in common use for manufactured products of all descriptions. Note the similarity (and differences) with the presentation of technical information about objects of natural form (Fig. 5.31 at the end of this chapter).

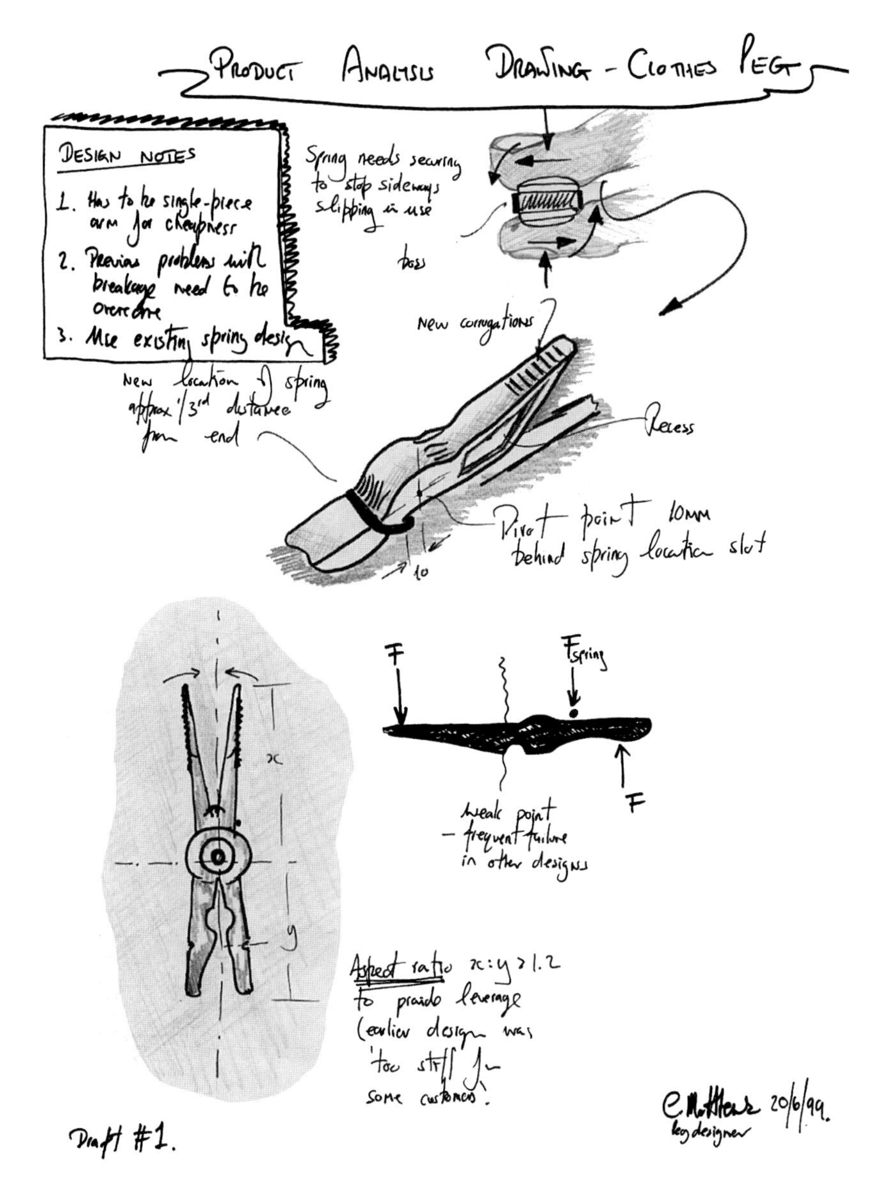

Fig. 5.4 A product analysis drawing

Product 'size-steps' diagrams

Many products and machines of all types are designed and manufactured in a range of size steps to match the requirements of the market. It is not only physical size that varies; for example, machines may vary in power output (or input), torque capability, speed, and similar function-related characteristics. Technically, size-steps are usually determined by a series of preferred 'R' numbers which provides the designer with a step size to work to.

The difficulty

Because of the many characteristics that change with the step size, it is difficult to show all the changes in a single diagram. Figure 5.5 shows the most common method, i.e. illustration line drawings showing the physical size steps accompanied by a multi-axis chart showing the main performance-related characteristics. Note the following features:

- the range of six product step sizes covered by the chart;
- the overlapping of performance ranges between step sizes;
- the cross-reference to a preferred number series via a step factor, in this case shown as Ψ.

Step sizes for a range of gear pumps

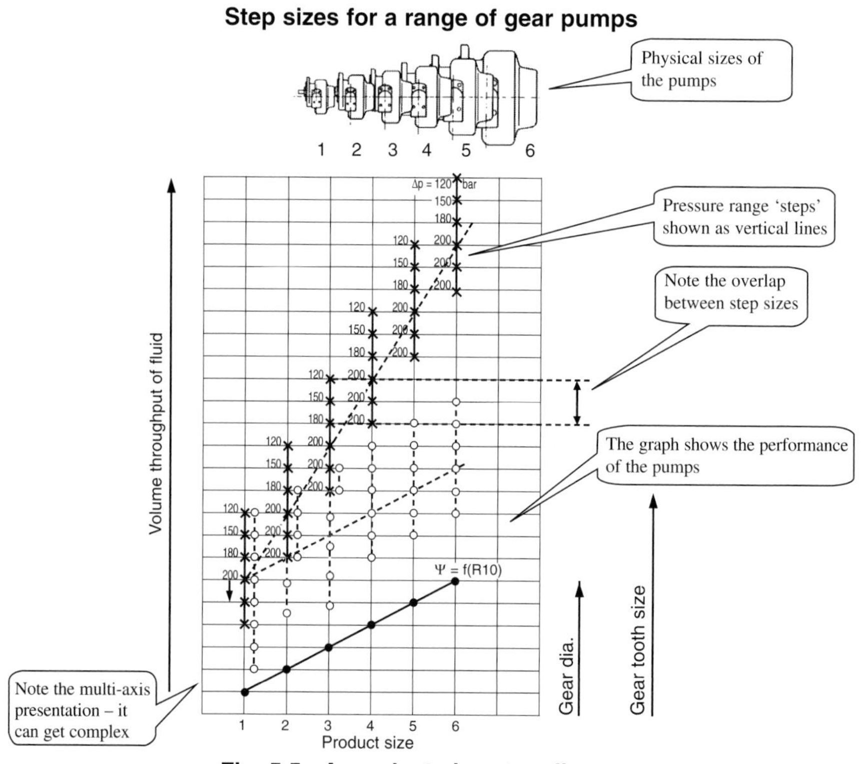

Fig. 5.5 A product size-step diagram

Dimension-only size-step diagrams

For static components such as structures, casings, etc., it is often only necessary to show the variation in physical dimensions resulting from product-range size steps. Figure 5.6 shows examples for a gearbox and an airliner. The illustrations are 'to scale' and show only the relative 'envelope' dimensions (no constructional detail is provided). Such illustrations can be shown in either 2-D or 3-D form.

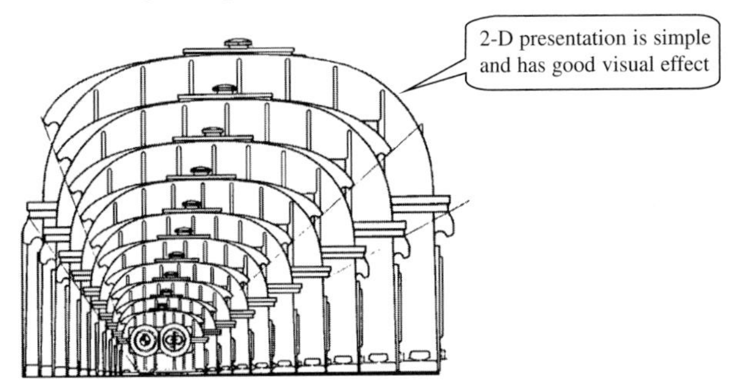

A range of short-haul airliner sizes

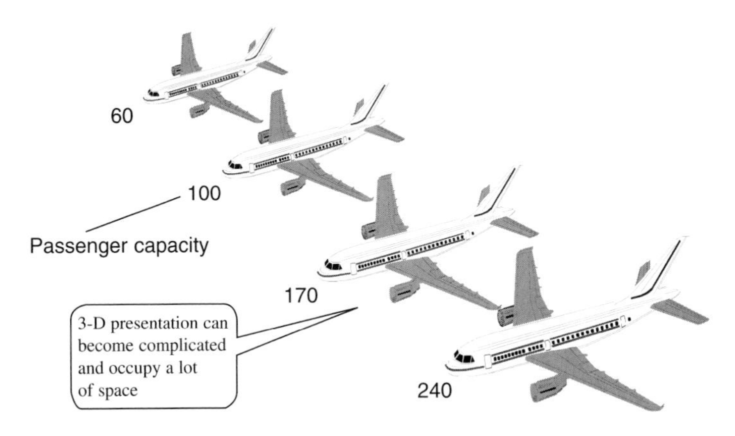

Fig. 5.6 'Dimension-only' product size-step diagrams

Combined performance and design feature drawings

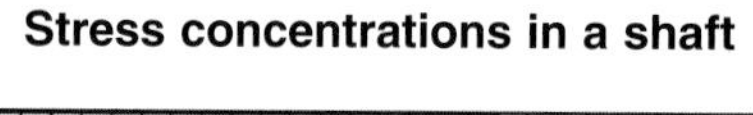

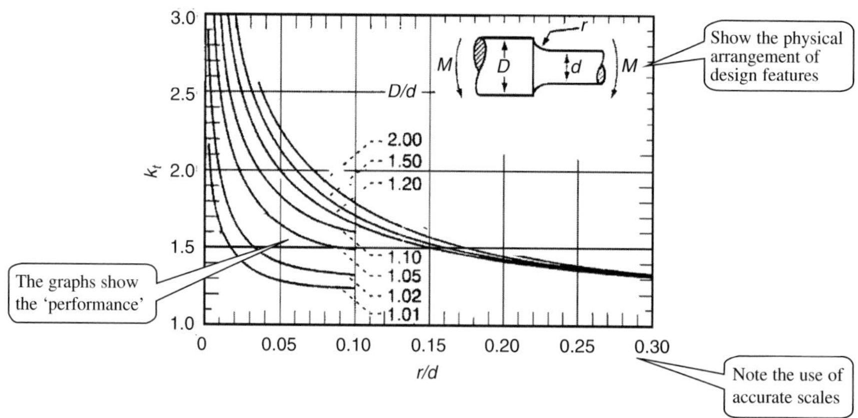

Fig. 5.7 A combined performance/design feature drawing

Individual performance characteristics and design features can be shown together on a combined drawing, such as that shown in Fig. 5.7. This type of presentation is restricted to simple components such as shafts, couplings, gears, bearings, etc., often grouped under the generic term of *machine elements*. Common factors are:

- a set of characteristic curves, shown on a traditional 2-axis line graph;
- essential dimensions and loading information about the component;
- accurate scaling of the horizontal and vertical axes as the chart contains accurate design data rather than general guidance information.

Component layout sketches

These are used to show the physical layout of the components in a machine or piece of equipment containing moving (normally rotating) parts. It is a symbolic diagram, using conventional graphic symbols to represent the various components and their interconnections. Figure 5.8 shows this type of sketch for a mechanical gearbox: note the symbols used and their meaning. The main purpose of such sketches is to explore design variants (i.e. different layouts for the gear trains, shafts, and couplings) rather than to convey any detailed technical or performance information about the design.

A gearbox layout

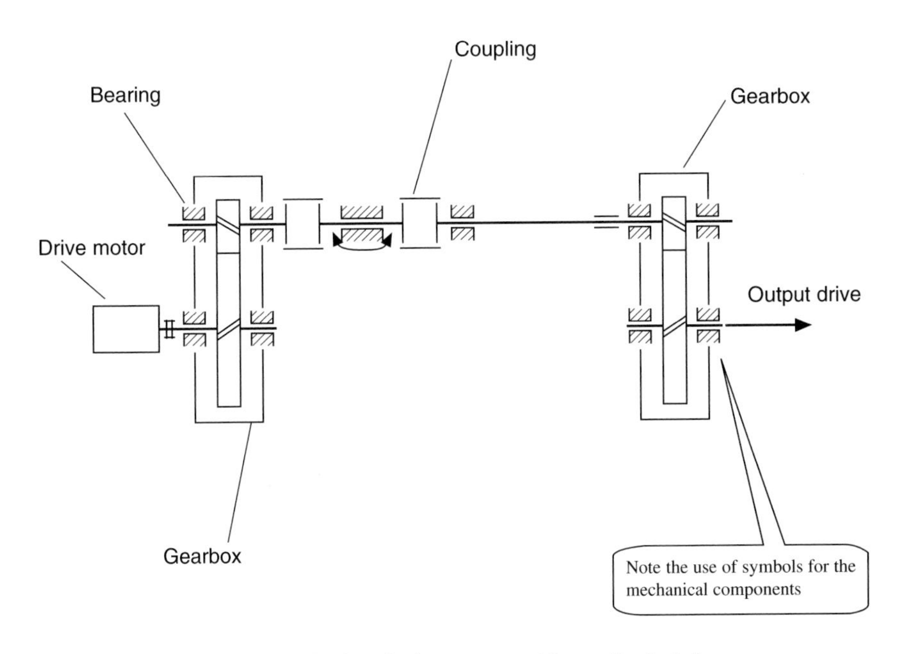

Fig. 5.8 A simple 'component layout' sketch

Comparison 'concept' sketches

These are sometimes known as 'pictosketches'. Their only purpose is to illustrate a comparison between two or more designs. The features are:

- unscaled, inaccurate sketches, intended 'for guidance only';
- explanatory text that illustrates direct comparisons between the same design features in different designs;
- simplified sketches with shading and colour to illustrate the comparison points.

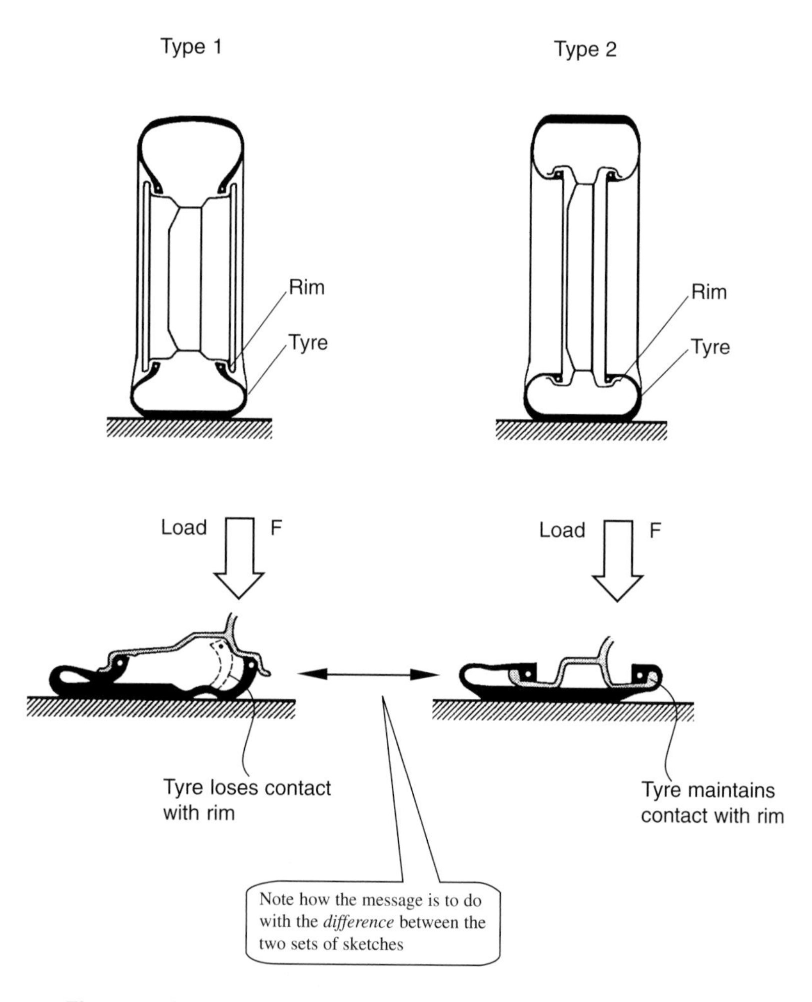

Fig. 5.9 Comparison concept sketches ('pictosketches')

Semi-detailed schematics

Schematics are a family of drawings which make only a partial attempt to represent reality. They provide a simplified view of a technical object or system by using a combination of drawing and symbols. Semi-detailed schematics are particularly useful in portraying engineering components. Their features are:

- limited technical detail about the object;
- very basic labelling (major sub-components only);
- no scale or dimensions, but the drawing is correct in basic proportion;
- no manufacturing details.

Figure 5.10 shows a typical example for a sliding/linkage mechanism. Note the internal detail shown, qualifying the figure as a semi-detailed schematic.

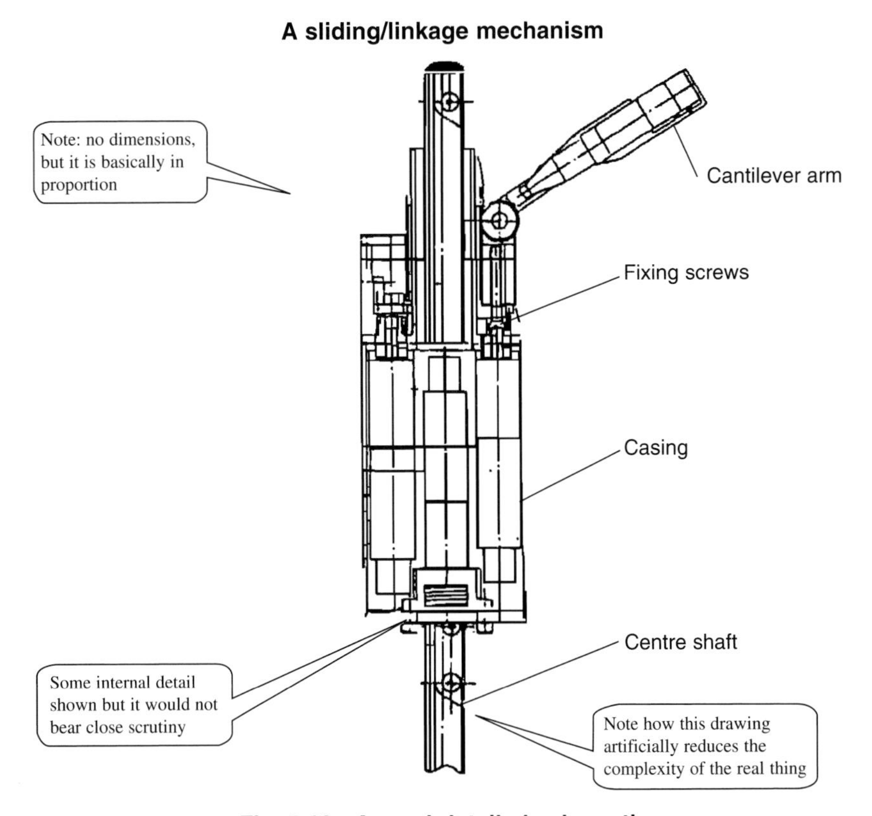

Fig. 5.10 A semi-detailed schematic

Useful reference

http://infosys.kingston.ac.uk/ISSchool/Research/d.fischer/App-lll.htm

Semi-pictorial illustrations

Semi-pictorial illustrations are often used in preference to schematic diagrams when they are intended for a non-technical readership. They are pictorial in that they use simple 'pictures' instead of technical-looking drawings, but they also contain explanatory text. This text usually containings information about function, i.e. how the item works. Note the features in Fig. 5.11:

- external views only, no internal detail;
- the drawing is pictorial, i.e. the drawings are strictly non-technical, and out of scale;
- some explanatory text is included.

Most illustrations which are classified as pictorial display these three features. Remember that pictorial means 'picture of' rather than 'accurate representation of'.

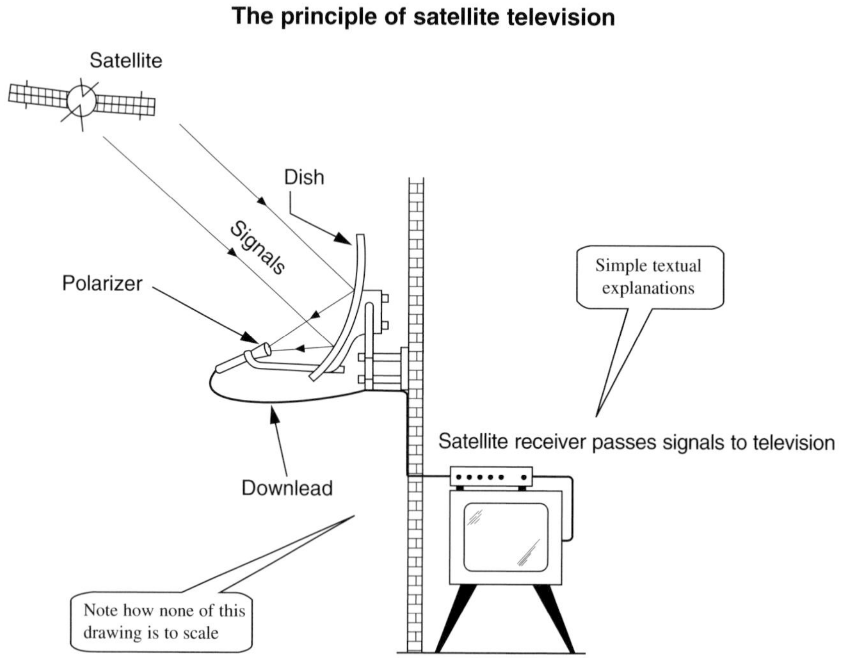

The principle of satellite television

Satellite
Dish
Signals
Polarizer
Simple textual explanations
Satellite receiver passes signals to television
Downlead
Note how none of this drawing is to scale

Fig. 5.11 A semi-pictorial illustration (simple)

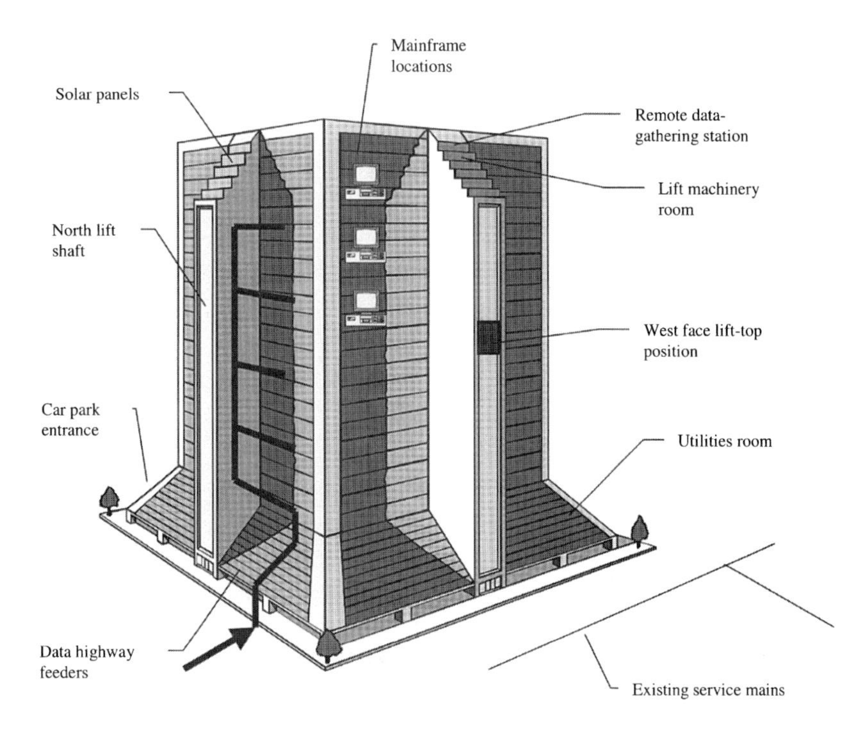

Fig. 5.12 A semi-pictorial illustration (complex)

Figure 5.12 shows a more complex type of semi-pictorial illustration. The principle is the same; it is a picture of an object (a building) but this time it is extensively labelled. Note how the labelling predominantly provides explanation of the internal contents of the building. This is an example of a much higher level of detailing. In common with Fig. 5.11, there are still no scale or dimensions and no attempt is made to accurately represent the shape or layout of the building's internal floors, etc.

Spatial schematics

These show the spatial relationship of multiple items (see Fig. 5.13). Note how there is no attempt to show any technical details of the oil platforms; the only purpose is to show the distances between them and how they are linked by oil pipelines. Spatial schematics are very common in:

- electrical/communication disciplines;
- transport/infrastructure network analysis;
- production engineering (factory layouts, etc.).

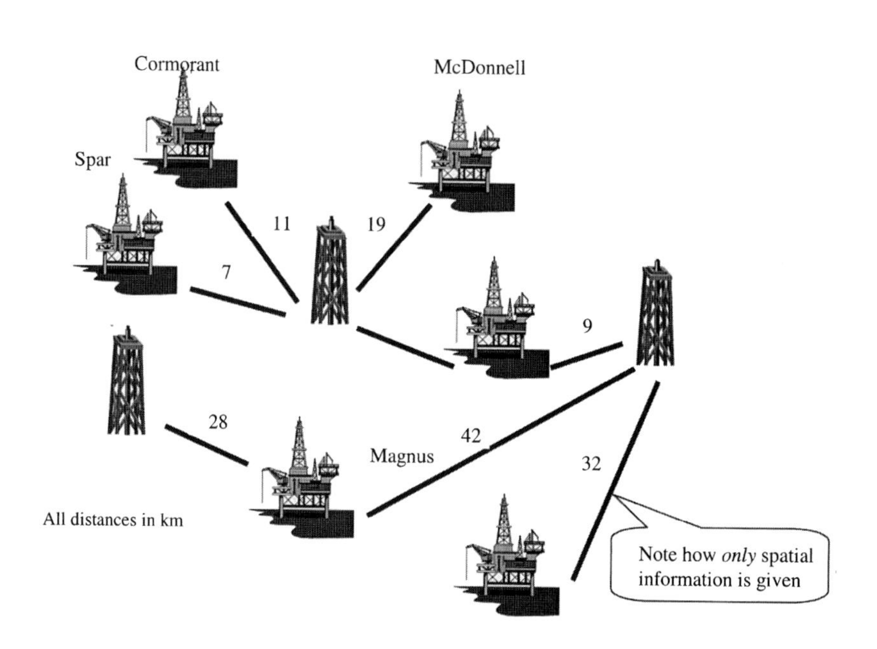

Fig. 5.13 A spatial schematic

Detailed pictorials

These are pure pictures. Detailed pictorial illustrations are normally targeted at non-technical and casual readers and are a common way to convey basic technical information in newspapers, journals, and popular magazines. Although greatly simplified, they are nevertheless still a valid way to present technical information. Indeed, people will often be attracted to look at detailed pictorial illustrations, whereas the same information presented in dry technical format would be ignored.

FOUR FEATURES OF DETAILED PICTORIAL ILLUSTRATIONS

- Large attention-grabbing title.
- A few (no more than four) short blocks of explanatory text, spaced around the illustration.
- 'Side panels' showing related general knowledge information using smaller illustrations and a single sentence of text.
- Pretty colours and shading.

Figure 5.14 shows a typical detailed pictorial which follows the above guidelines. There is a wide variety of this type of illustration in use, but all follow these general principles.

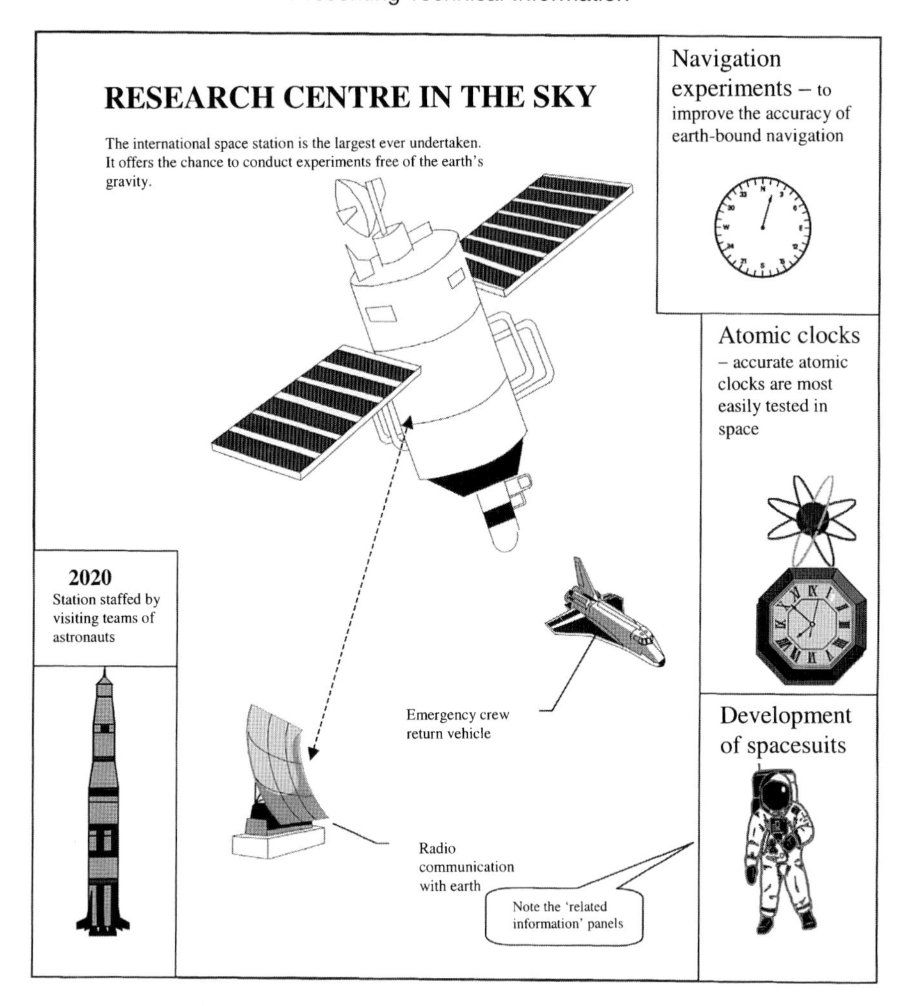

Fig. 5.14 A detailed pictorial illustration

Operational sketches

Operational sketches show how things work. They can be 2-D or 3-D views; normally isometric projection (see Fig. 5.15) and show outlines only rather than a lot of surface or construction detail. Features are:

- all the components are labelled, with the labels also referring to the function of individual components;
- motion arrows shown for all rotating/reciprocating/oscillating components;
- no dimensions or manufacturing information.

Operational sketches can vary from the simple to the very complex. Complex machinery and assemblies (electro-hydraulic items, engines, etc.) can often only be shown in simplified 2-D view, otherwise they become too difficult to understand.

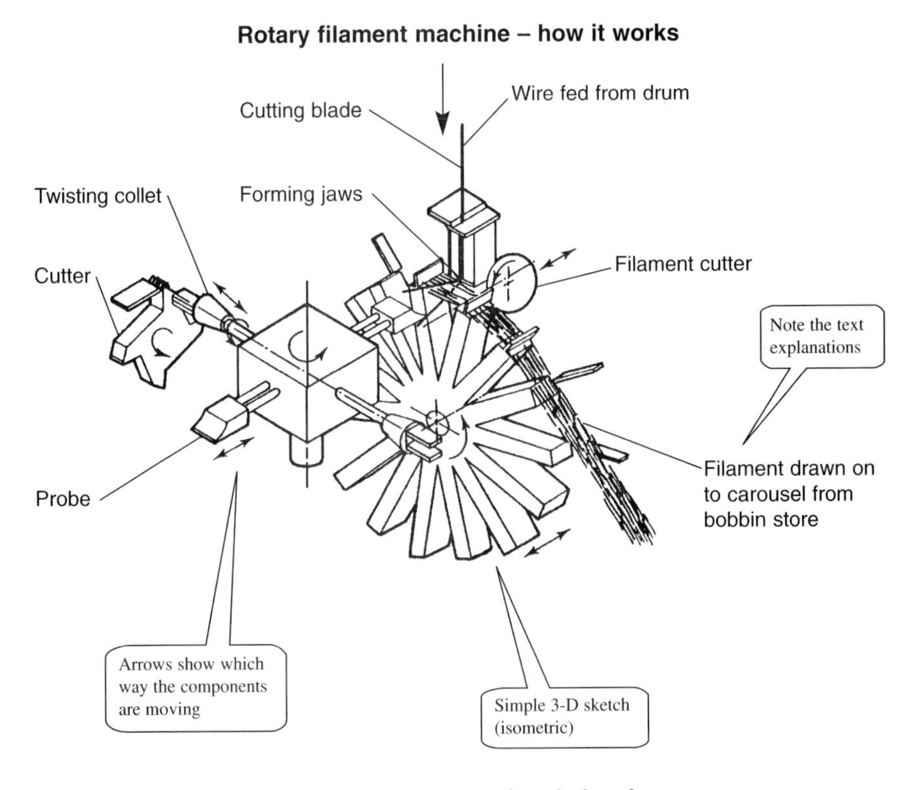

Fig. 5.15 An operational sketch

Evolution sketches

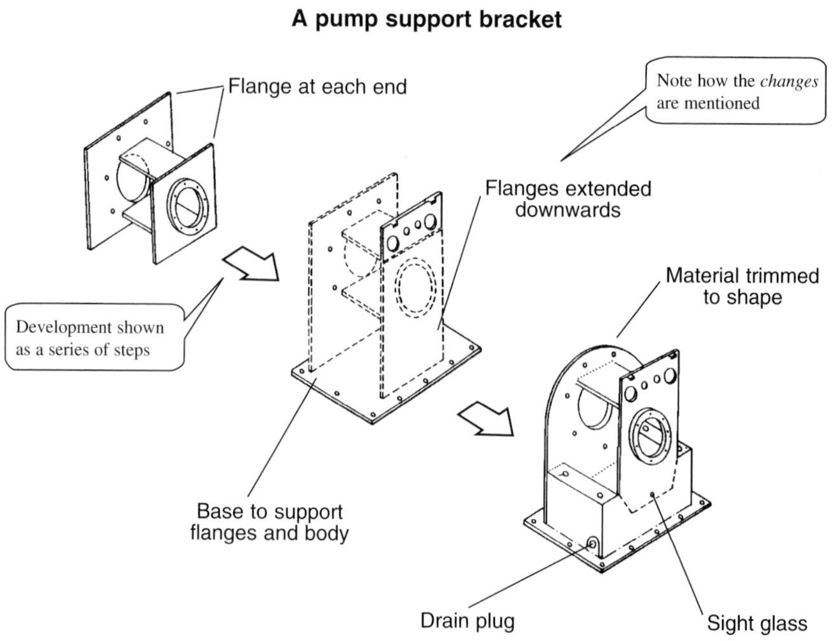

Fig. 5.16 An evolution sketch

Evolution sketches are found in both the conceptual and practical stage of the technical design process. Figure 5.16 shows a typical version from the practical stage.

EVOLUTION SKETCHES ARE ABOUT:

- Developing ideas;
- Making changes;
- Showing features of design changes.

Evolution sketches are also a way of explaining design variants. Richer, more advanced design ideas can generally be better developed in 3-D sketches than the simplified 2-D form seen in Fig. 4.6.

Ergonomic diagrams

'Ergonomics' or 'human factors engineering' is the study of the relationship between humans and technical objects. It is an important part of the technical design process, providing input data to several subsequent stages of design activity. There are two basic types of ergonomic diagrams:

- those showing ergonomic considerations;
- scaled ergonomic drawings.

Ergonomic 'consideration' sketches

These show the key ergonomic 'issues' of a technical design. Figure 5.17 shows an example. Note how issues such as visibility, noise levels, etc. are identified and shown as being important. Such drawings may be symbolic, schematic, or semi-pictorial, depending on how complex the technical design is.

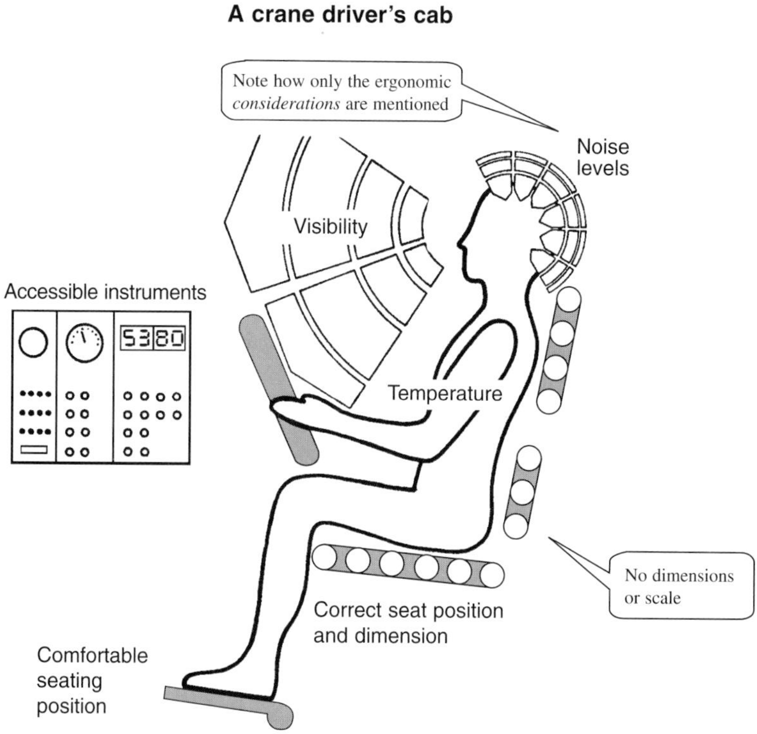

A crane driver's cab

Fig. 5.17 An ergonomic 'consideration' sketch

Scaled ergonomic drawings

These are the next stage on from the 'consideration' sketches. Drawn to scale, they show the detailed 'fit' of a design to the human body, i.e. they illustrate the exact nature of the user's requirements. Ergonomic drawings rely on the availability of anthropometric data, i.e. on the size of humans. Standard data are available which are normally applicable to 90 percent of the population. Figure 5.18 shows an example for an office chair. It is convention to use standard orthographic projection for the drawings. Several variants are available which show, for instance, the design in several stages of orientation or adjustment.

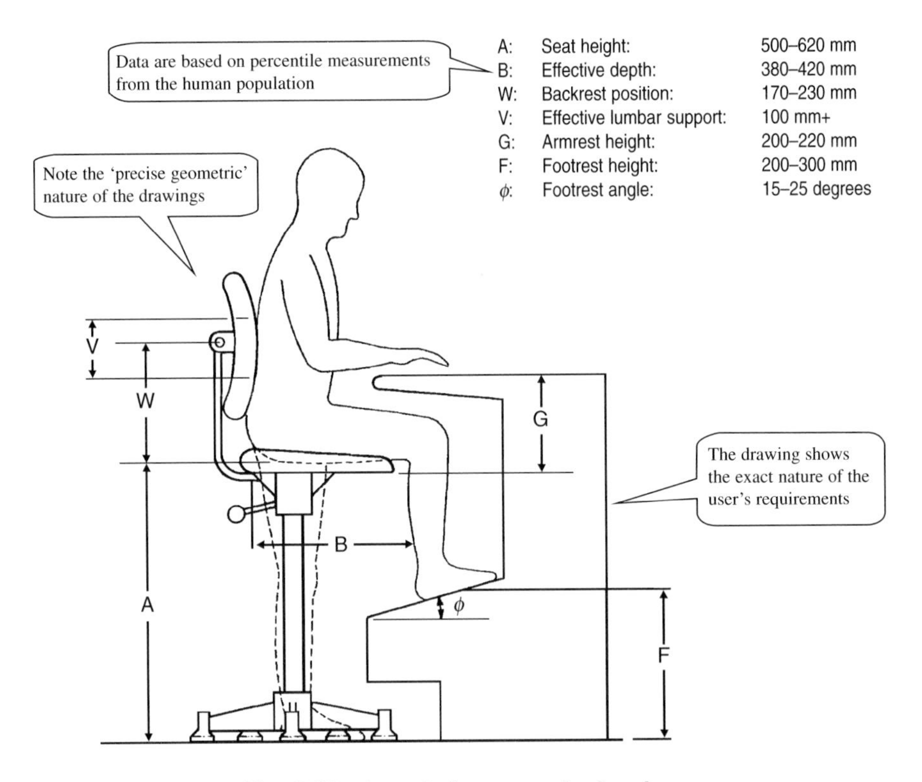

Fig. 5.18 A scaled ergonomic drawing

Useful references

http://members.aol.com/canyonergo/Forum.htm
BS 3044: 1995 *Guide to Ergonomic Principles in the Design and Selection of Office Furniture*, The British Standards Institution.

Stand-alone illustrations

Stand-alone illustrations are used in the later stages of the design process and are often used in instruction manuals. Their objective is to be easily understood by a very wide range of people who speak different languages. A typical example is given in Fig. 5.19. Note the features:

- an illustration which is self-explanatory without accompanying text;
- accuracy and attention to detail;
- pictorial view, normally isometric with 30° or 45° projection angle.

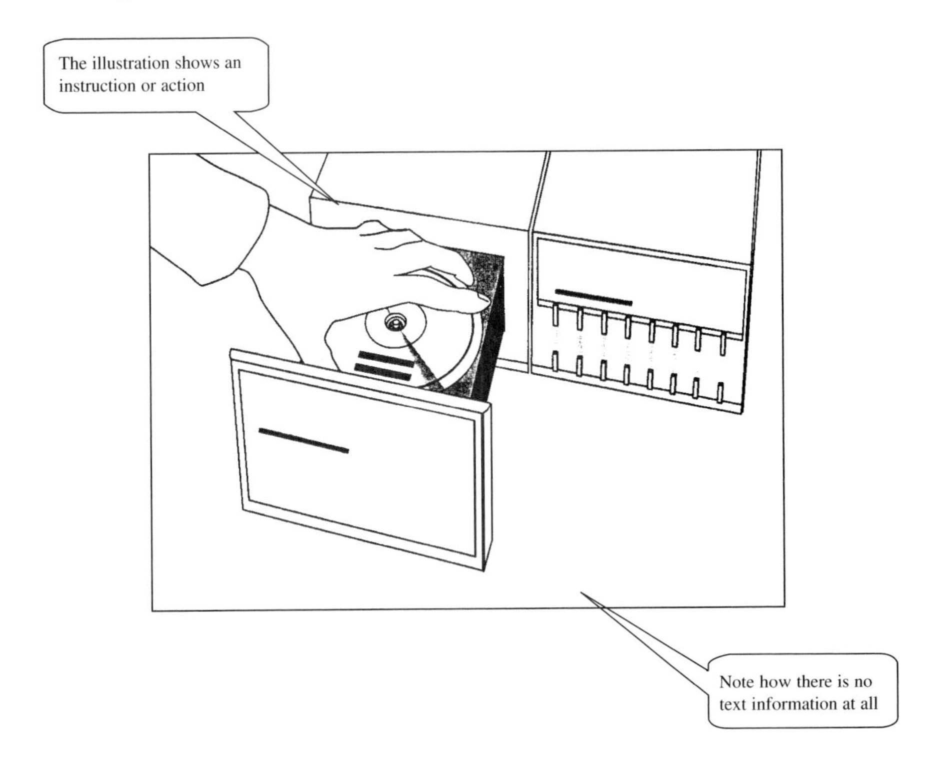

Fig. 5.19 A 'stand-alone' illustration

Product dimension and shipping information

This is a specific type of technical presentation which provides information related to the transport and shipping of a completed product. It is typically included in the final documentation package that accompanies a product as part of its sale. Figure 5.20 shows an example for a large engineering product. Note the features:

- 'outside' dimensions for deciding the packing case size and transport method;
- outline views only, no technical detail;
- net and gross weights;
- broad details of technical function for reference purposes.

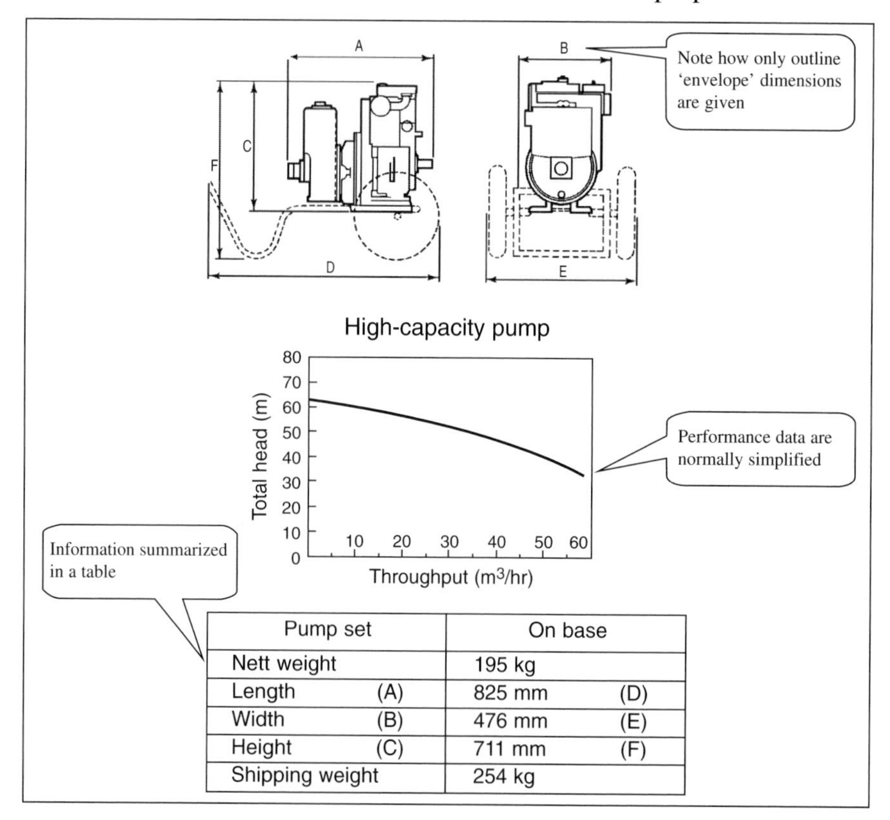

Pump set		On base	
Nett weight		195 kg	
Length	(A)	825 mm	(D)
Width	(B)	476 mm	(E)
Height	(C)	711 mm	(F)
Shipping weight		254 kg	

Fig. 5.20 Product dimension and shipping information

This information is generally formatted on to a single sheet for convenience. Various parts of the outline view may be shown dotted (as in Fig. 5.20) and it is common to summarize the dimensions showing the actual size in an accompanying table rather than on the drawing.

'Picto' product data

This slightly unusual form of presentation appears in product literature such as sales brochures and outline specification sheets. It is mainly intended for the use of purchasers, some of whom will not be fully conversant with reading technical drawings.

> **'PICTO' PRODUCT DATA – PRESENTATION FEATURES**
>
> - Simple symbols, often enclosed in a 'block'.
> - Symbols arranged in rows or columns.
> - One title per symbol, plus a line or two of explanatory text.

There is no universal standard on the symbols used. Manufacturers generally use their own designs, perhaps borrowed from their product catalogues or spares lists.

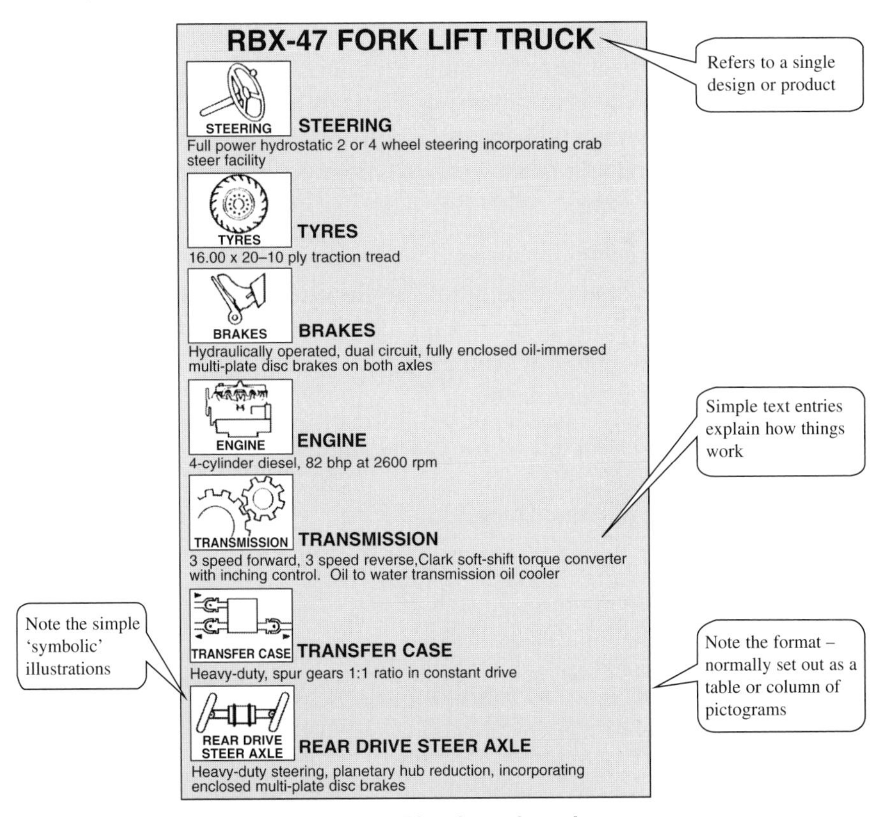

Fig. 5.21 'Picto' product data

Sales brochure presentations

Technical presentations in advertising and sales brochures follow a specific and well-known format. This material is intended purely for the purposes of persuading people to like a product (as well as probably buying it) often at the expense of providing proper technical details. The presentation can be considered as comprising two parts: illustrations and text.

Illustrations

Note the 'overall view' type of illustration shown in Fig. 5.22. It is an isometric projection and shown with some of the operating features in use (in this case the document feeder). Line illustrations are rarely suitable for sales brochure presentations; it is better to use a photograph view to show colour and surface texture. Photographs are also better than drawings at showing style. The illustrations are nearly always labelled, but with the number of labels kept to a minimum so as not to clutter the visual appeal of the illustration.

It is rare for only a single illustration to be shown on each page of a sales brochure; expect to see one or two smaller 'thumbnail' photographs on the same page. Figure 5.22 gives a good example of this. Note how the small illustrations in the bottom corners are only marginally relevant to the main product. This is a common sales presentation technique and is meant to add 'balance' to the page.

Text

Text is normally formatted in columns for clarity. The idea of clarity and ease of assimilation is crucial in sales presentations. Text has to be capable of being assimilated almost at a glance if the presentation is to fulfil its purpose. Other text features are:

- simple bold typeface used for sub-headings;
- technical specification terms written in italics to differentiate them from qualitative parts of the text.

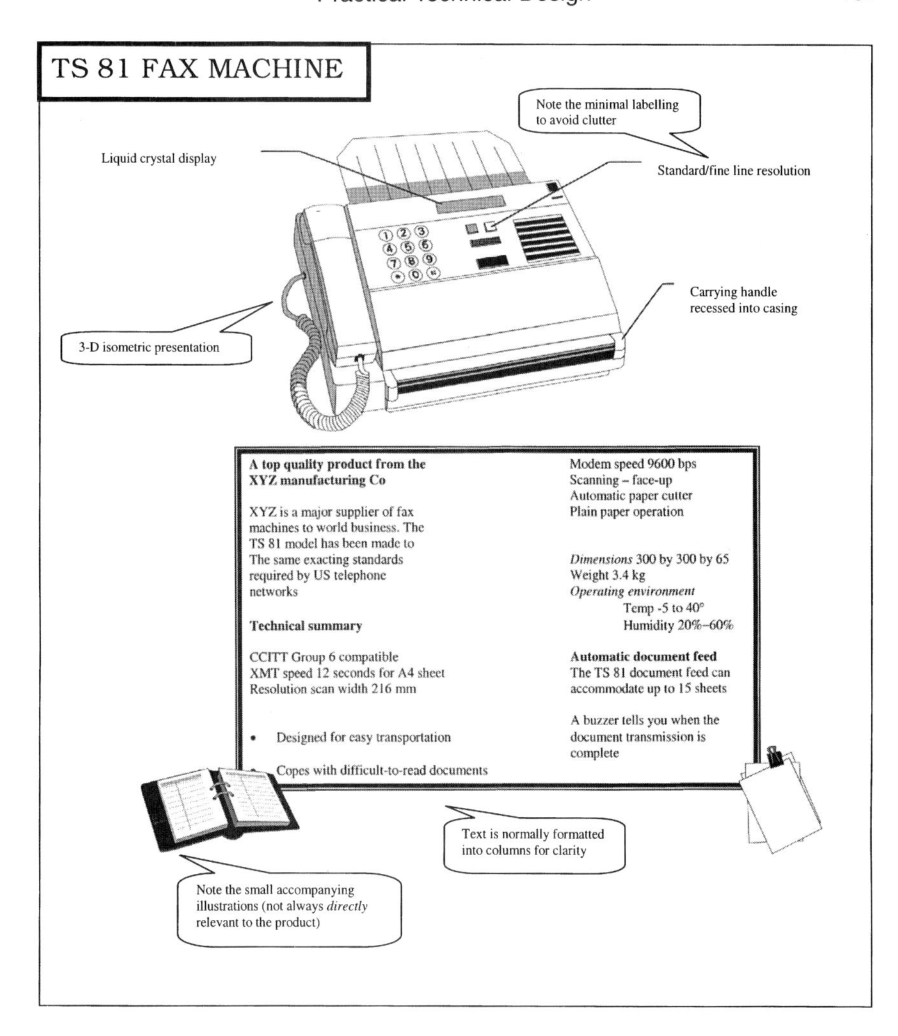

Fig. 5.22 An example of 'sales brochure' presentation

Product purchase data sheets

Product purchase data sheets are a more technical version of the product data found in advertising and sales brochures. They have the status of illustrated technical data sheets and are used in technical catalogues and by buyers to select products from a range. As with product shipping data, they are generally formatted on a single page.

THERE ARE THREE PARTS TO A PRODUCT PURCHASE DATA SHEET

- Overall 'envelope' dimensions.
- A table of dimension and weight data.
- Some information on materials of construction.

Look how the data given in Fig. 5.23 are a little thin on technical details. They are only a sample, and not intended to be a full description of the technical performance of the product. Note also the format of the data sheet:

- the outline drawing at the top;
- dimensional and weight information in the centre, in tabular format;
- materials information in abridged form.

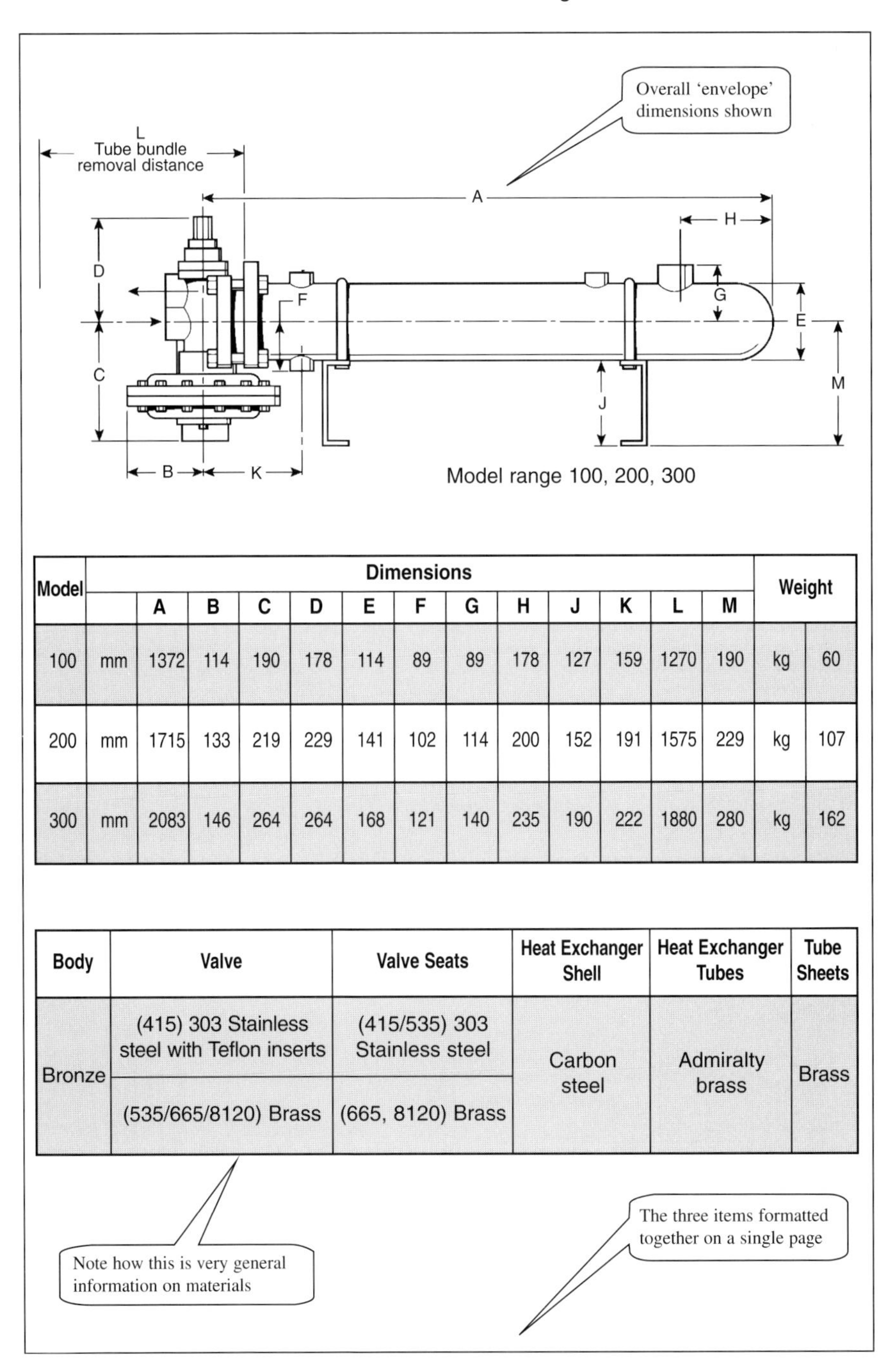

Model	Dimensions													Weight	
		A	B	C	D	E	F	G	H	J	K	L	M		
100	mm	1372	114	190	178	114	89	89	178	127	159	1270	190	kg	60
200	mm	1715	133	219	229	141	102	114	200	152	191	1575	229	kg	107
300	mm	2083	146	264	264	168	121	140	235	190	222	1880	280	kg	162

Body	Valve	Valve Seats	Heat Exchanger Shell	Heat Exchanger Tubes	Tube Sheets
Bronze	(415) 303 Stainless steel with Teflon inserts	(415/535) 303 Stainless steel	Carbon steel	Admiralty brass	Brass
	(535/665/8120) Brass	(665, 8120) Brass			

Fig. 5.23 A 'product purchase' data sheet

Product installation diagrams

Product installation information is included in the document package when a technical product is supplied to a purchaser. It shows how a technical product 'fits together' with other parts of the system that are there already. Figure 5.24 shows the installation diagram for the heat exchanger in Fig. 5.23. Note the use of colour or shading to differentiate the new product from the existing system. Other features are:

- pipework fittings, etc. are traditionally shown as 2-D line drawings rather then just schematic 'lines';
- arrows show the directions of fluid flow, or movement, and how they match up with the flow, etc. in the new product to be installed;
- simple labelling of the item names only; no explanation of technical detail or function.

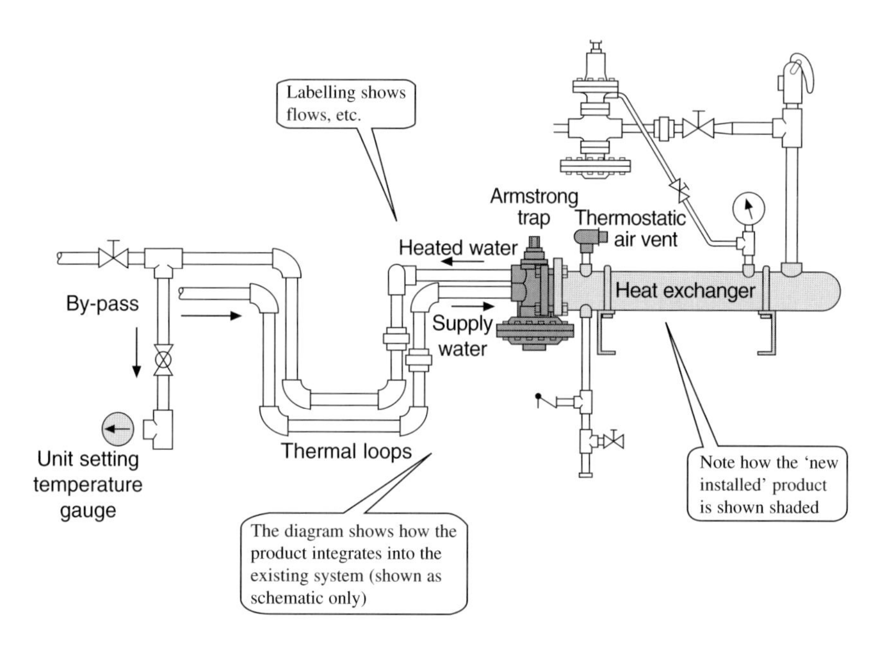

Fig. 5.24 A 'product installation' diagram

Postscript: showing style

The subject of style can be controversial in some technical disciplines. The traditionalist view is that style is relegated to second place behind functionality and that it is the reserve of the artistic disciplines, rather than those which are purely technology based. This view is changing. The discipline of Industrial Design, for example, is about interleaving the techniques of art and technology in the creation of products which then have aesthetic appeal as well as functional efficiency.

So what is style?

Style is:

- line
- colour
- proportion
- texture
- light and shadow
- spatial relationships

All of these elements of style are inherent in technical objects and so it is necessary to use them in illustrations and presentations that represent these objects, at least if you want to do the job properly. So, the message here is:

STYLE

is

an integral part of presenting technical information
(not an add-on or a separate subject)

Types of style

The technical world is awash with different kinds of style. The hard conventions of orthographic projection and technical drawing have produced an established style of presentation. Computerized techniques such as CAD and 3-D modelling have done the same thing, albeit a style which is still developing. Within these two extremes lie a myriad of different ways to show technical information, each portraying different information and messages about the object that is being represented.

Precise geometric style

Figures 5.25 and 5.26 show examples of drawings that reflect a precise geometric style. Although they are not accurate orthographic views, the style and character of the presentation suggests a high level of geometric precision. This infers the nature of the product itself as being of precise technical design with a highly refined shape and accurately fitting components.

PRECISE GEOMETRIC STYLE IS ABOUT:

- a single, constant line thickness;
- an uncluttered image, often against the background of some kind of geometric grid;
- no colours, hatching, or surface texture.

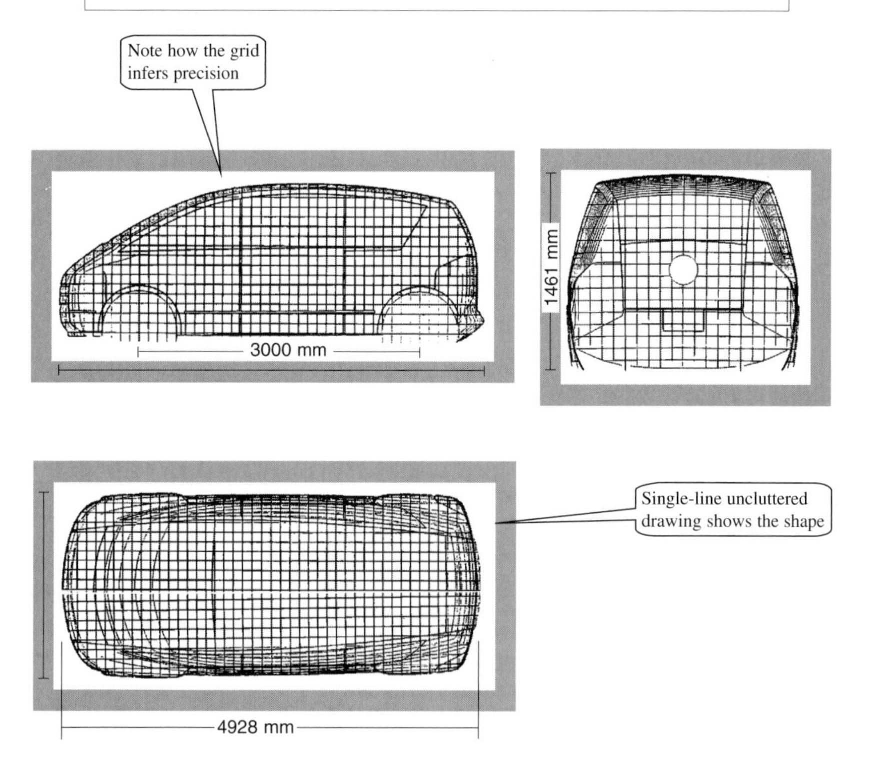

Fig. 5.25 Precise geometric style – in 2-D

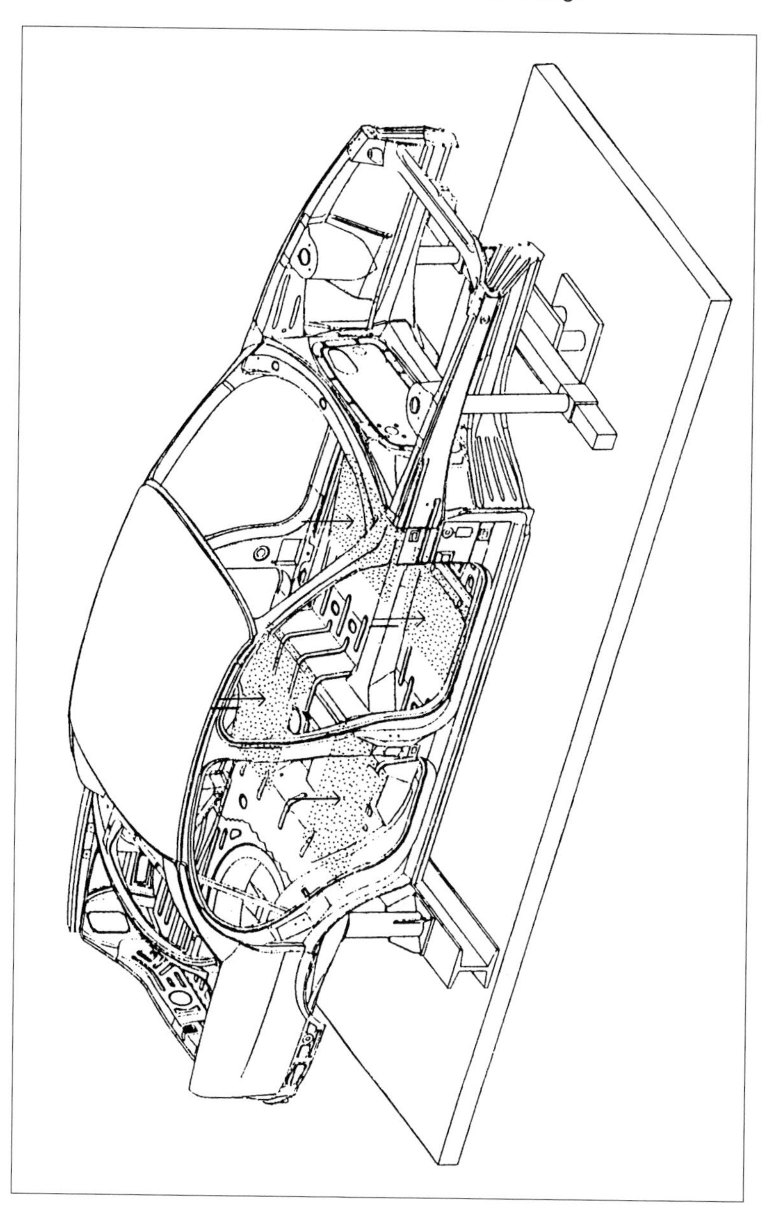

Fig. 5.26 Precise geometric style – in 3-D

Nominal aesthetic style

This is actually a mixture of styles: a conventional engineering approach laced with a minimal amount of aesthetics. The objective is to convey both technical and geometrical rigour but not at the exclusion of aesthetic appeal. Figure 5.27 shows a typical example: look at the almost minimalist approach of this technical drawing, using lots of white space to emphasize the flowing lines of the roof shape. There is sufficient conventional technical detail (the vertical spacing of the floors and the use of 'scale' people, etc.) to convey the necessary engineering and size aspects. The presentation of the drawing is, however, still aesthetic. A technical drawing of this nature benefits from the generally 'minimalist' approach. Figure 5.28 shows another example, this time designed for a greater visual impact.

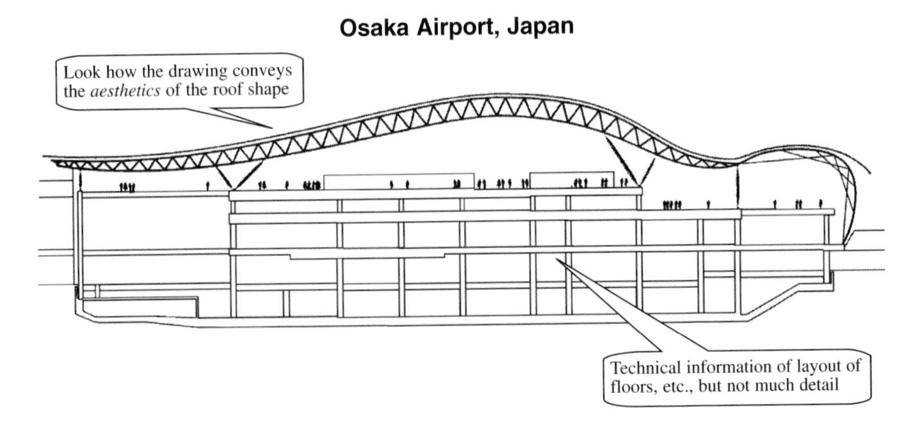

Fig. 5.27 An example of 'nominal aesthetic' style

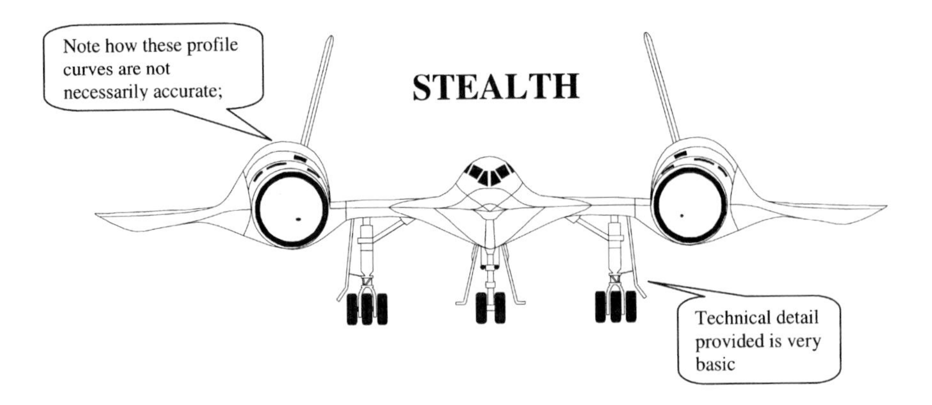

Fig. 5.28 Nominal aesthetic style – another example

International style

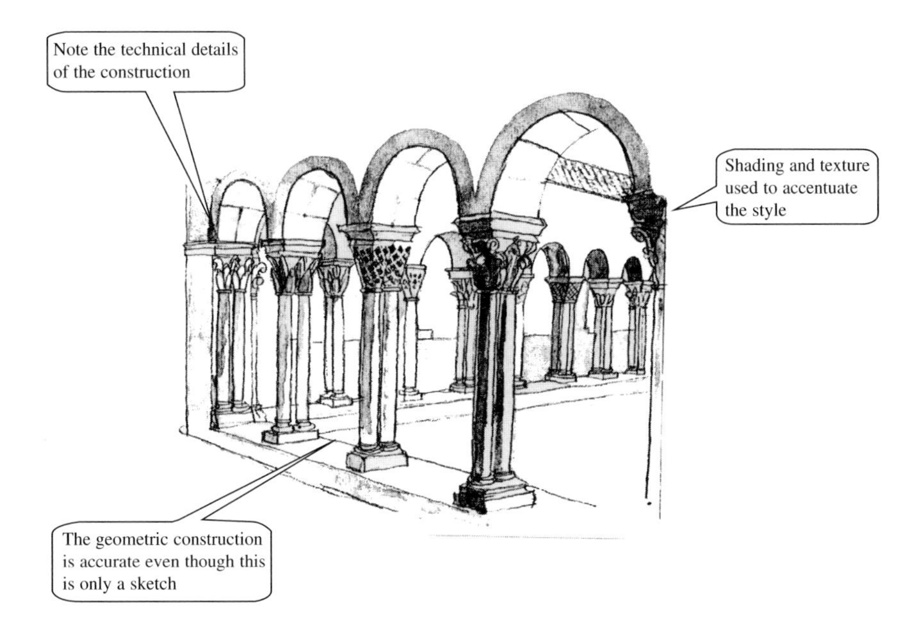

Fig. 5.29 An example of distinctive 'international style'

The term 'international style' originated in the field of architecture, with its roots in the European trends of the 1920s and 1930s. Over time, the usage of the term has spread outwards to apply to any architectural, industrial or technical design that is geometric, but which possesses the quality of asymmetry. Technical presentation in subjects as diverse as town planning, civil engineering and, of course, pure architectural design often reflect this type of style. Figure 5.29 shows a typical example. Note the key points:

- rigid geometric construction (isometric in this case) with accurate verticals, horizontals, and curves;
- shading and surface texture added to selected areas; this conveys character;
- portrayal of constructional detail (look at the tops of the columns).

Abstract style

In art, the term 'abstract' refers to forms that have no direct relevance to reality and are abstruse and difficult to understand. In the technical world it carries a looser definition, inferring that it is an image or technical representation that has been altered in some way to provide a slant on the way that technical details are seen. There is an almost endless variety of abstract ways to convey technical information. The three most common uses are shown in Figs 5.30 (a) to (c).

Figure 5.30(a) is a 3-D 'wireframe' drawing. In their most crude form, wireframe drawings are more of a sketch than an accurate drawing, unlike 3-D finite element displays, etc. which can look similar. Their purpose is simply to convey something about the 3-D shape of a technical object, without attempting to specify close detail.

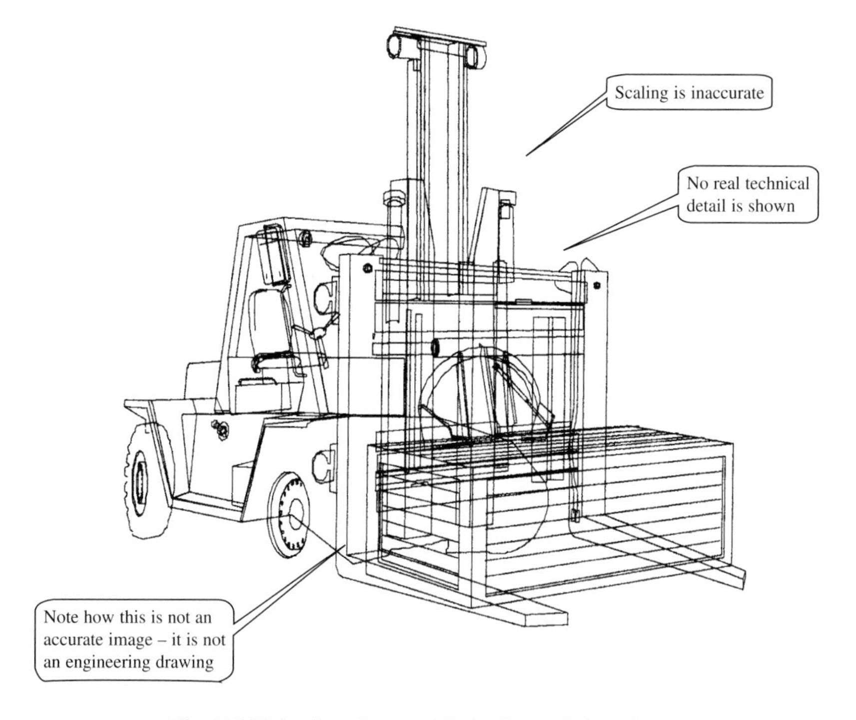

Fig. 5.30(a) An abstract 'wireframe' drawing

Figure 5.30(b) is a different type; this is abstract because of the way in which it shows multiple views of a single object. The representation (in this case an orthographic view) is often quite accurate – certainly better than in wireframe drawings.

Figure 5.30(c) is a 'geometric' abstract drawing, almost a silhouette. These are useful in conveying firm visual messages about geometric style, mainly in architectural objects rather than detailed technical components.

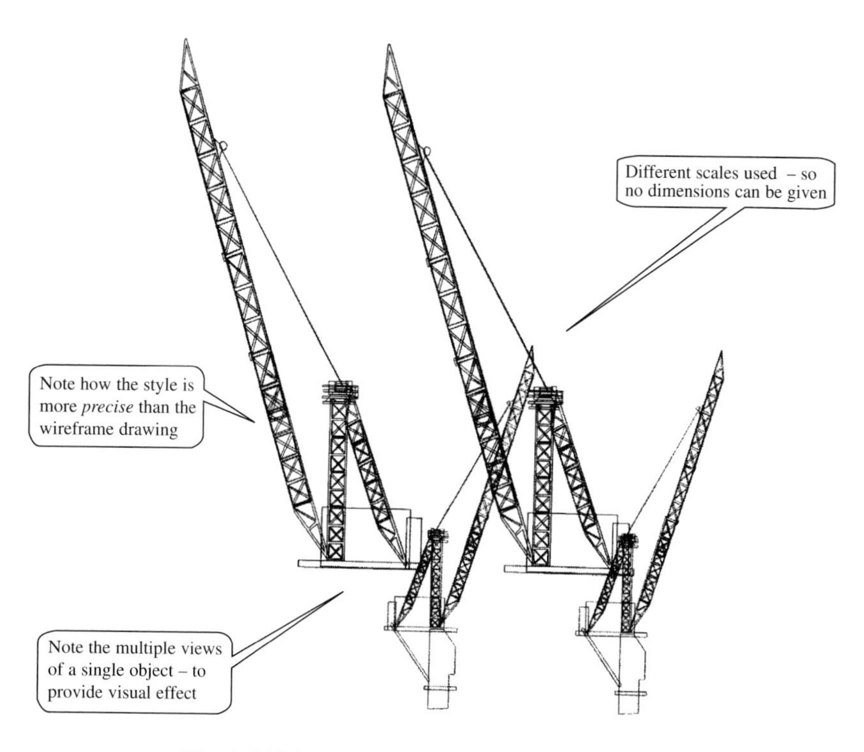

Fig. 5.30(b) A 'multiple view' abstract drawing

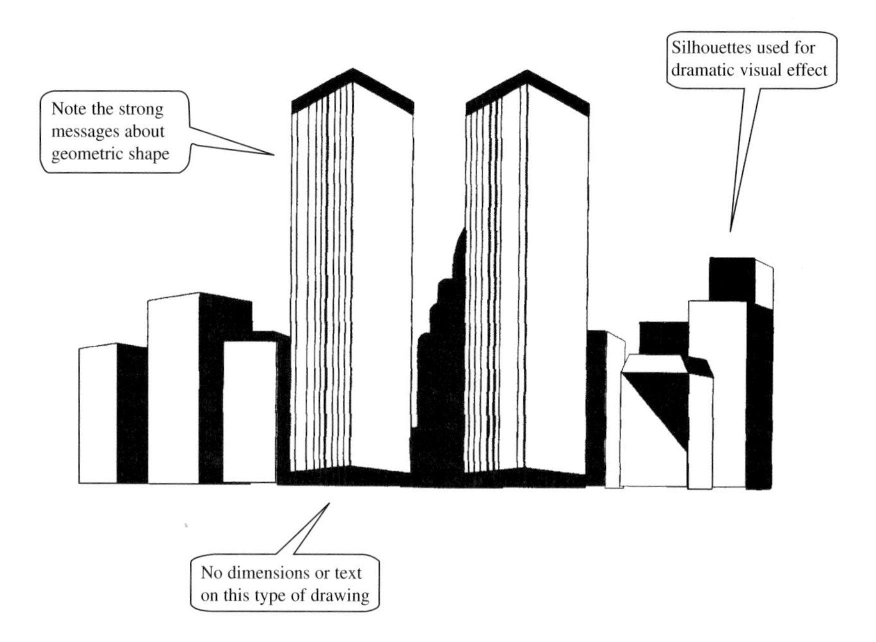

Fig. 5.30(c) A 'geometric abstract' drawing

'Life geometry' style

It is not always easy to relate the presentation of technical information to the world of living things, because technical objects are conventionally thought of as being inanimate; followers of their own technical rules rather than those of nature. In practice, this is untrue; all technical objects follow the rules of physics, which are the rules of the natural world. Reflect further and it starts to become clear that a lot of engineering objects, designs, and systems mirror those found in nature, both in principles and physical geometry and form. The idea of a 'life geometry' style, therefore, has relevance to many technical disciplines. Figure 5.31 shows a classic example – the logarithmic spiral of the Nautilus mollusc. Note the common features.

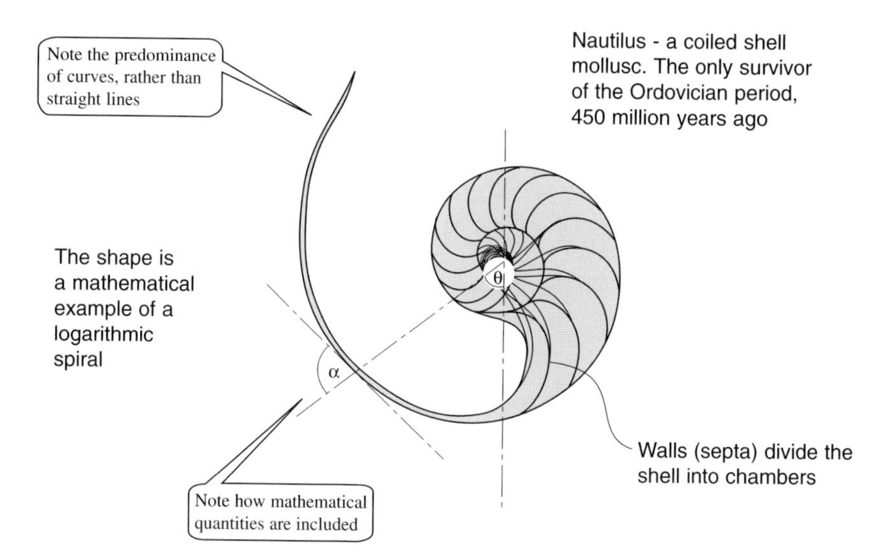

Fig. 5.31 'Life geometry' style drawing – the Nautilus

FEATURES OF 'LIFE GEOMETRY' STYLE

- Freehand illustrations (because there are not many straight lines in nature).
- Lots of curves and angles.
- Approximate dimensions.
- The use (or inference) of mathematical equations (because these do reflect nature).

And then: curious topologies

The design of any object is concerned with, among other things, the shape and other quantitative geometrical features of its surface. There is another much less well-known type of geometry – that of non-quantitative geometry which is studied using the discipline of *topology*. Topology is to do with the geometry of place and position rather than the conventional (Euclidean) subject which is interested in lengths, distances, angles, and suchlike. You can think of topology, therefore, as being the study of the position and relation of the inside, outside, and parts of an object without regard to their shape or size.

Some bizarre results

Topology can produce some bizarre, often surreal, results to the design process. They are not real designs, but not unreal either. Figure 5.32 shows three examples – living proof that you can draw these 'topological' items; but could you make them?

The bottle with no inside (and no outside)

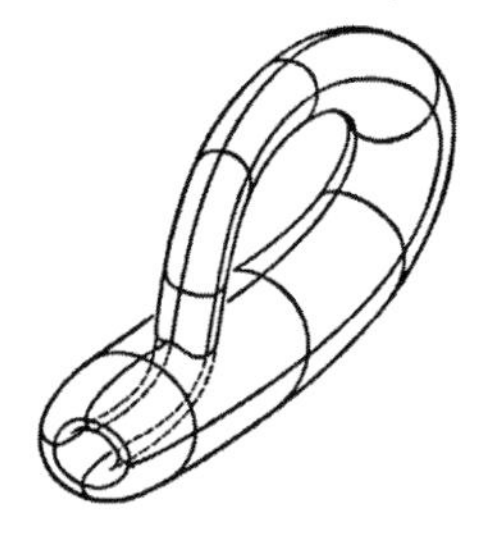

The hole through a hole in a hole

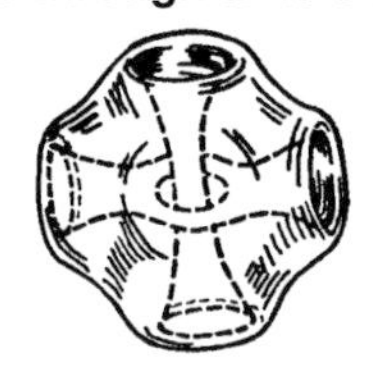

The discontinuous pretzel?

Fig. 5.32 Bizarre topologies – sketches from the edge of reality

Chapter 6

Reliability Information

Introduction – reliability

The topic of reliability appears in many different technical disciplines. Terms such as 'design for reliability', 'reliability assessment', 'risk analysis', and similar have become commonplace when discussing technical objects and systems. Reliability is not always an easy concept to deal with. It consists of combinations of robust mathematical theories, coupled with a more practical empirical approach about the real-world behaviour of complex objects and systems.

The presentation of technical information about reliability is also not easy. Whilst the subject is built around a structure of rigid mathematics it is necessary to combine this with a more user-friendly pictorial approach that non-specialists will understand. This chapter shows the more practical types of presentation that are in common use across the technical disciplines.

The confusion diagram

This is a common, no-doubt well-intentioned, method of trying to explain reliability (in any technical discipline). The general idea is that subjects and topics that impinge upon the reliability of a design are shown in the rather random type of bubble diagram shown in Fig. 6.1. Such diagrams are too general, too random, and do not work. Their only merit is as a diagram which causes confusion. Look at some of the features:

- different sizes of 'bubble', with no real message as to what they mean;
- various sizes and styles of typeface: again, you have to guess the significance;
- all manner of different types of links trying to show the form of interrelationships between the bubbles.

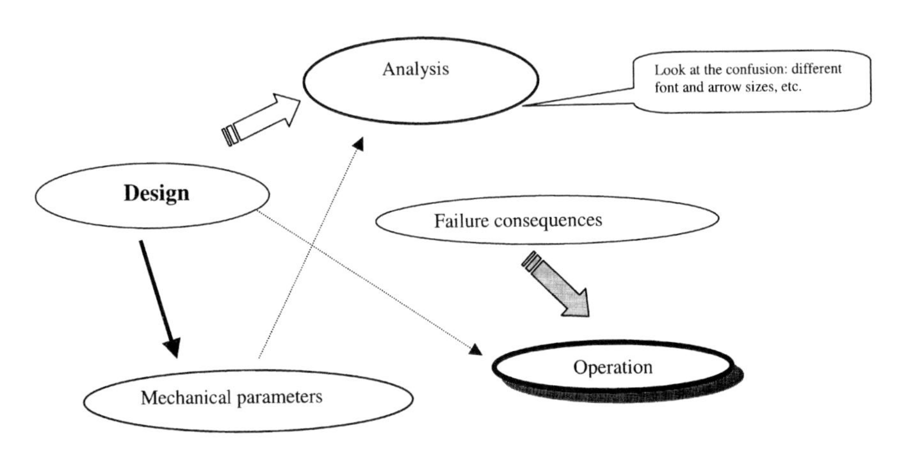

Fig. 6.1 Reliability – the confusion diagram

Although diagrams like Fig. 6.1 may not be actually *untrue*, they do little to convey the ideas and concepts of reliability to their reader. Furthermore, they can never be accurate enough to be a viable method of presenting technical information about reliability. So, do not use them.

Related-factors diagrams

A central concept in the subject of reliability is that of interrelationships between 'factors'. The simplest situation is demonstrated in the presentation method shown in Fig. 6.2(a), which shows three factors that influence the structural integrity of an engineering component. Note how the use of a triangle, with each 'factor' represented as a separate apex, infers that all the factors have equal importance, i.e. there is no general hierarchy. These simple triangle diagrams tend to be used to show the existence of reliability-related factors for objects or systems that are inherently simple.

In contrast, Fig. 6.2(b) is hierarchical. It shows related factors, but in a way which infers that there is some hierarchy behind the relationship, i.e. some factors are bigger, more important, or more directly related to the issue of reliability than others. This diagram should be used when there is a hierarchy and when the number of related factors you want to show is larger than three.

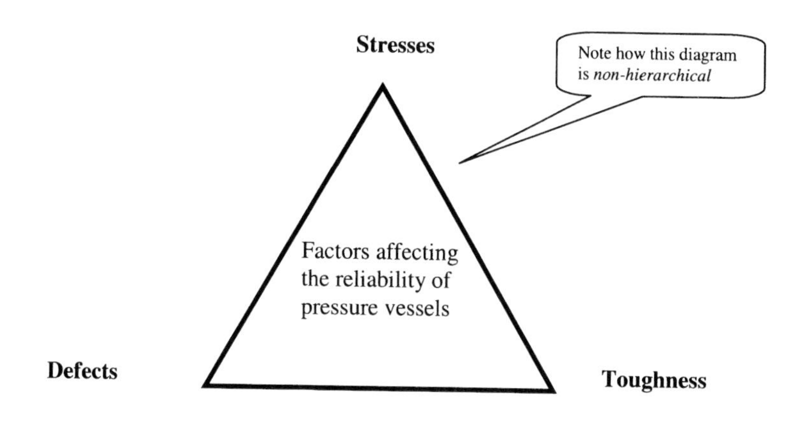

Fig. 6.2(a) A related-factors diagram (with no hierarchy)

Reliability and risk assessment in industry

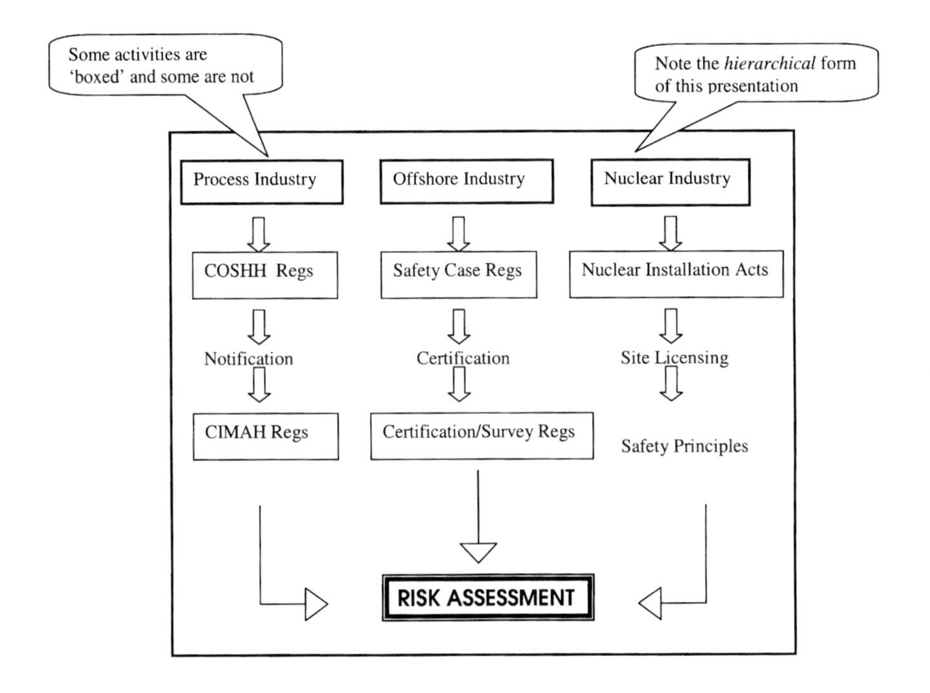

Fig. 6.2(b) A related-factors diagram (with hierarchy)

Related-factors diagrams – pictorial

'Related-factors' diagrams can also be shown pictorially, as in Fig. 6.3. The principle is similar to that in Fig. 6.2(b) except that the information in text boxes is replaced by a 'pictogram' or symbol. This is a useful type of presentation to use for reliability-based subjects because it forms an attractive contrast to the dry mathematical equations that often follow it. Note three general features of the pictorial method:

- pictogram images are not as direct and precise as text, so the diagram will have a more 'general' feel to it;
- it is easy to clutter the diagram by using too many pictograms; any more than seven or eight and all the different images will start to confuse the readers, not inform them.
- pictorial diagrams can be made hierarchical (like Fig. 6.2(b)) or non-hierarchical; just adapt the layout to suit.

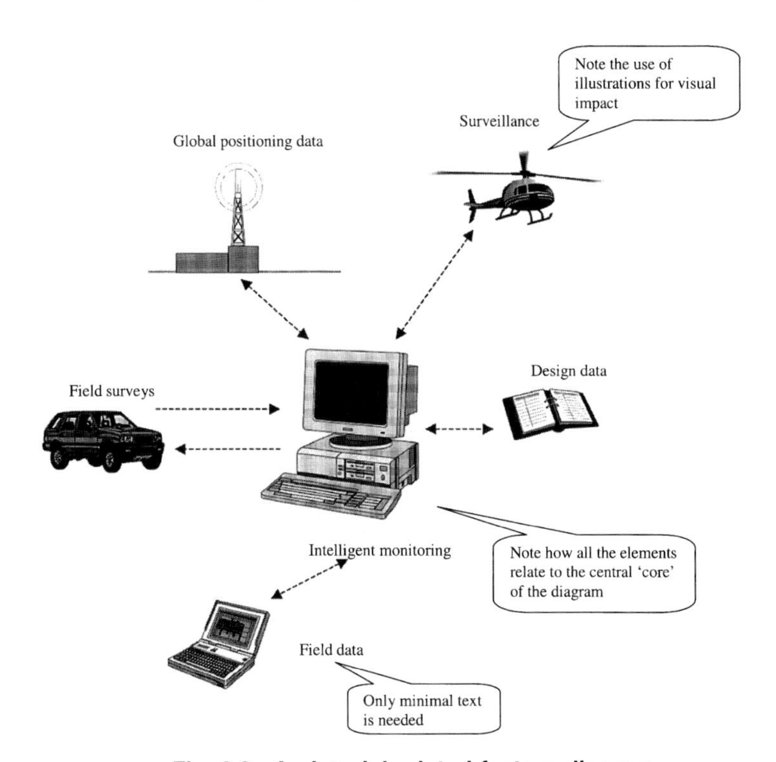

Fig. 6.3 A pictorial related-factors diagram

Useful reference

http://manta.cs.vt.edu/cs5224/ObjectDiagrams/ObjectDiagrams.html

Risk categorization

The assessment and categorization of risk is a common thread running through many technical disciplines. It also forms a part of various reliability-related techniques. Technical information about risk categorization is generally presented in one of two ways: (a) the risk versus consequence chart or (b) the network method.

Risk versus consequence charts

This is a straightforward 2-D chart linking the risk of failure with its consequences. Figure 6.4(a) shows an example. Note how combinations are given a quantitative (points) rating so that the chart can be adapted for use in all technical disciplines, as long as there is no need to show any complicated mathematical or statistical relationships.

The network method

This is a rather misleading title because the diagram does not actually show any representation of a network, as such. It is a way of displaying a complex set of risk or reliability relationships in a simplified way. Figure 6.4(b) shows an example. Look how quantitative values of risk or reliability are shown in the vertical direction. Note also the semi-pictorial 'feel' to the diagram, to prevent the mathematical relationships of the network becoming too oppressive.

Pressure vessel risks

	Risk of failure			Consequences of failure		
Points	*Age (yr)*	*Type records*	*Hydrotest records*	*Internal pressure*	*Class*	*Flange size*
3	>20	All cast		Design pressure	600 or unknown	>18"
2	10–20	Fabricated body			300	12" –18"
1	5–10	Cast body body/fabricated closure	No records		150	8" –12"
0	<5	All fabricated	Records available	Down rated		<8"

Fig. 6.4(a) A risk versus consequence chart

An approach to oilfield risk analysis and management

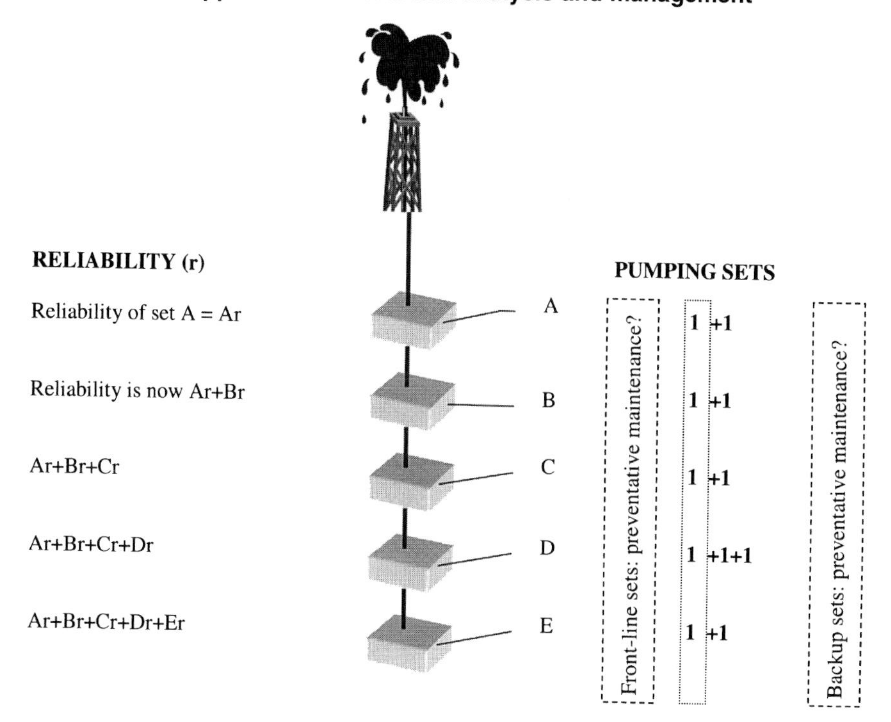

Fig. 6.4(b) Risk categorization – the 'network' method

Identification of critical equipment

In the engineering disciplines, reliability assessments are closely allied to the identification of which components of a product or system are critical to its function. This is relevant both at the design stage and the later operational stages during inspections, remnant life studies, and similar. The conventional way to present this kind of analysis is via the type of flowchart shown in Fig. 6.5. This allows a fairly easy visual assimilation of what is critical and what is not, without involving the reader in the (often subjective) technical decisions that lie behind the content of the chart. Note the other features:

- the use of standard flowchart symbols for clarity;
- the inclusion of several different decision criteria, i.e. duration of inspection/repair, feasibility of repair/replacement, etc. (the bottom half of Fig. 6.5).

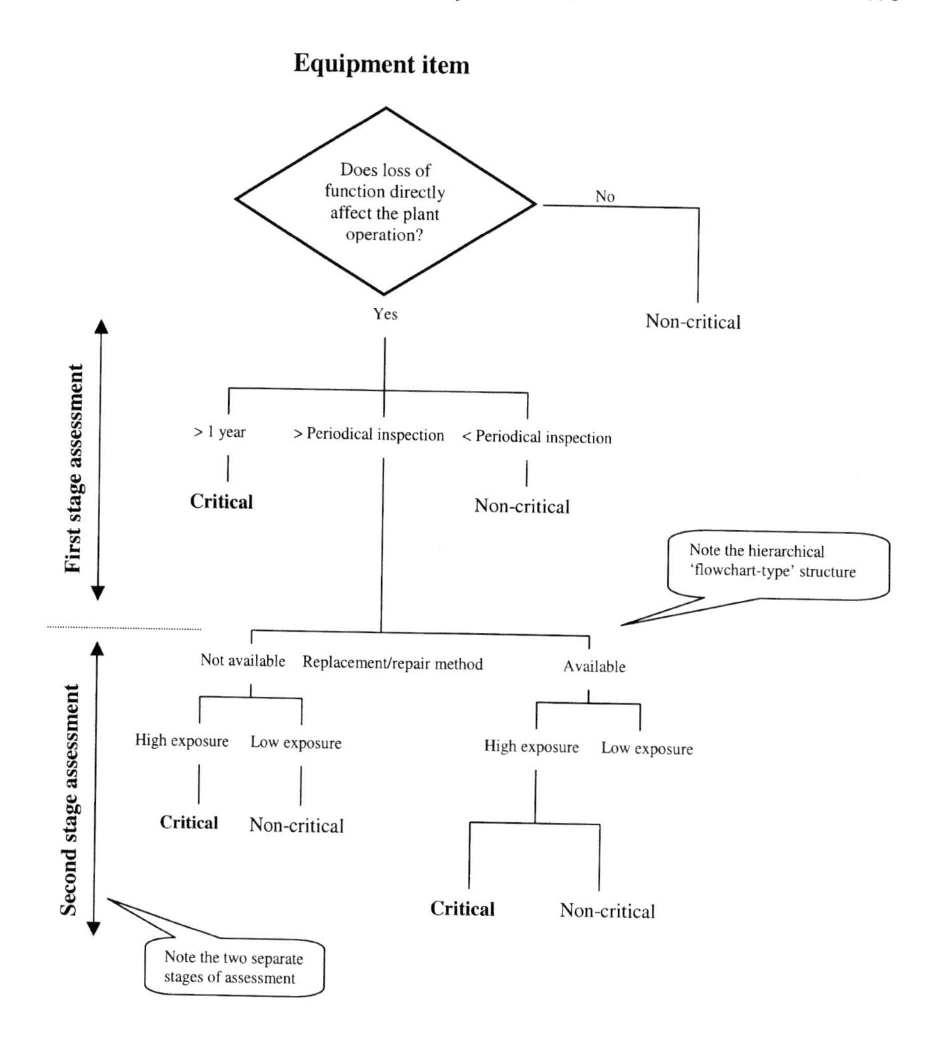

Fig. 6.5 Critical equipment identification chart

Reliability analysis flowcharts

The quantitative, step-by-step nature of reliability analysis means that flow charts can be used to show procedures and results. Figure 6.6 shows one common type of reliability analysis flowchart. It simulates the series of steps that comprise a reliability analysis of a system (in this case a process engineering application) and the way that various quantitative pieces of information are used within it. This style of analysis flowchart is useful in almost any technical discipline where reliability analysis forms part of the design process.

RELIABILITY ANALYSIS FLOWCHART – KEY FEATURES

- The flowchart is orientated vertically, with the sequence running from top to bottom.
- Quantitative bits of data such as Mean Time to Repair (MTTR), etc. are shown as information input using standard flowchart symbols.
- The flowchart is purely sequential, without nested or complex links.

Determination of test interval

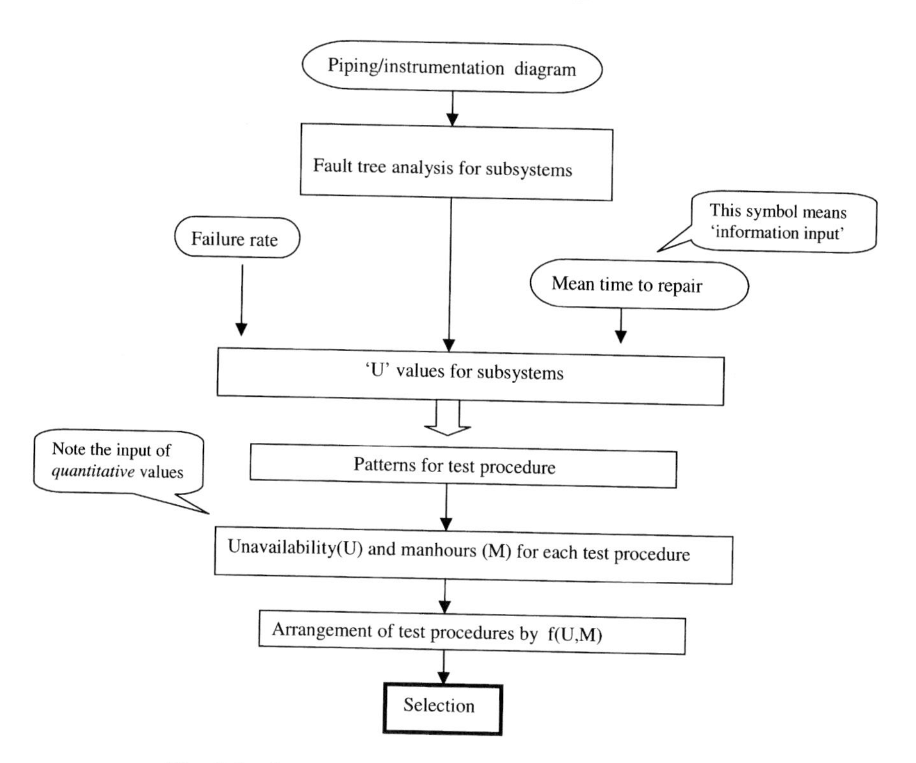

Fig. 6.6 One type of 'reliability analysis' flowchart

Fishbone diagrams

Formally termed 'reliability networks' (or something similar) these are more commonly referred to as 'fishbone diagrams' because they resemble the shape of a fishbone. Their purpose is to show the branched networks that are a feature of the subject of reliability, either in design or operation. Figure 6.7 shows a qualitative version of this type of presentation, i.e. where all the branches represent features or properties, rather than numerical data.

Fishbone diagrams (qualitative and quantitative types) follow the same general set of conventions.

FISHBONE DIAGRAMS – KEY FEATURES

- The branches run horizontally across the page from left to right.
- Key data entries are placed on the main spines of the fishbone, and are often enclosed in boxes for emphasis.
- The fishbone is drawn as single solid lines of constant thickness. Elaborate line formats would only confuse the message of the diagram.
- Arrowheads are not used. This is because the diagram is not strictly unidirectional (e.g. left to right), even though it is drawn that way.

Useful reference

http://www.powertechnic.com.au/AnalystArticle.htm

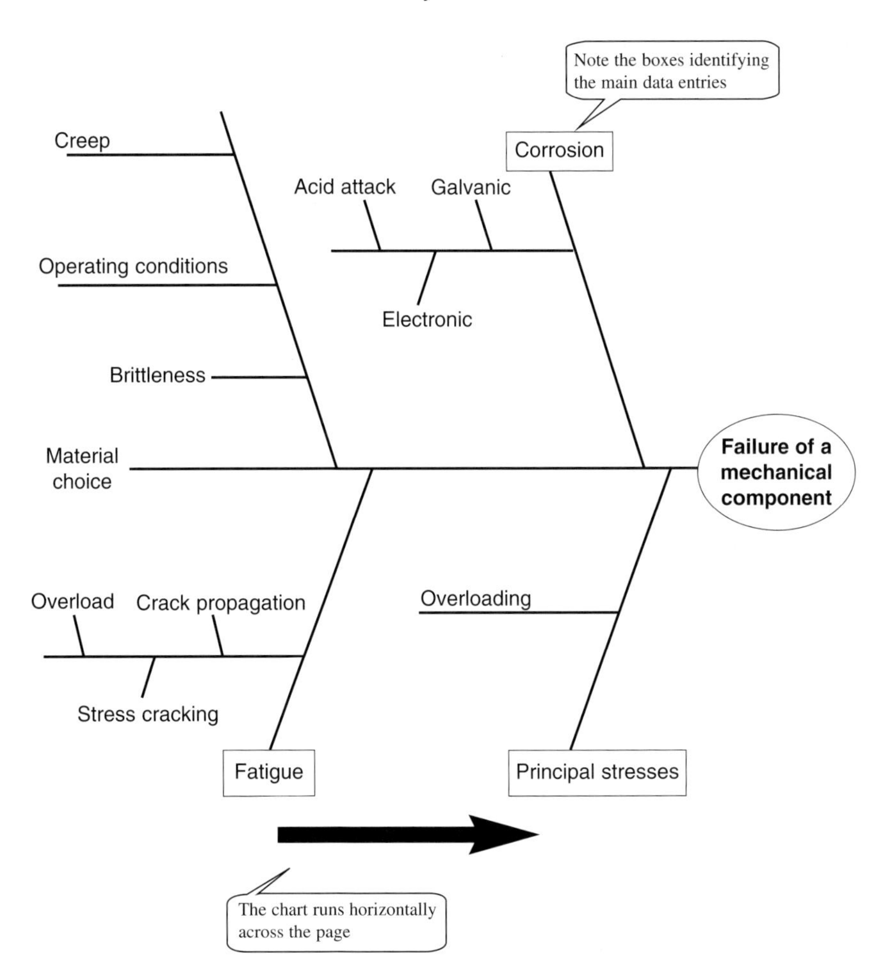

Fig. 6.7 A fishbone diagram

FMEA matrices

Failure mode effects analysis (FMEA) is a formal method of analysing the way that components of an assembly or system can fail and what the effect of that failure is. FMEA is a recognized technique in all the engineering fields, from simple manufactured items to complex process plants, aircraft, and similar. FMEA information is presented in the matrix form shown in Fig. 6.8 Such matrices can become very long and complex when applied, for example, to large process systems, but the principles remain the same. Note the essential features:

- functions of the product or system are shown in the rows of the matrix;
- failure modes are shown in the columns;
- a quantitative weighting system is used, enabling the results to be condensed down to a single quantitative value (usually expressed as a percentage) for comparison with other products or systems.

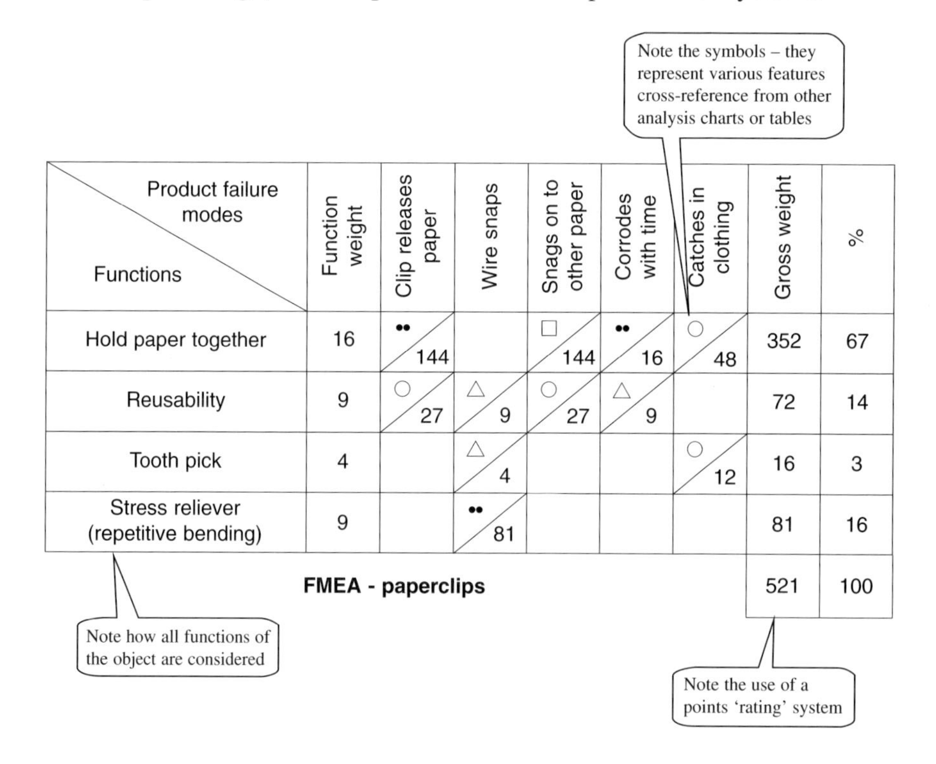

Functions \ Product failure modes	Function weight	Clip releases paper	Wire snaps	Snags on to other paper	Corrodes with time	Catches in clothing	Gross weight	%
Hold paper together	16	•• 144		□ 144	•• 16	○ 48	352	67
Reusability	9	○ 27	△ 9	○ 27	△ 9		72	14
Tooth pick	4		△ 4			○ 12	16	3
Stress reliever (repetitive bending)	9		•• 81				81	16
							521	100

Fig. 6.8 One type of FMEA matrix

Chapter 7

Mechanics – Statics

Introduction to statics – what is it?

Statics is the part of mechanics that deals with things that either do not move, or are only moving at constant velocity. In practice it is mainly concerned with structures. Primarily, statics is about the study of forces and the effects that they have on structures and the materials from which these structures are made. Statics is, therefore, a design discipline.

Statics – its technical information

Mathematical expressions are important in statics because they are a convenient and accurate way of expressing Newton's laws (which govern the subject) and analysing the practical problems and ideas that result. Statics is also, however, about geometry and spatial relationships – concepts which are better shown by the use of graphical presentations and diagrams. Diagrams in statics rely heavily on conventions. Fortunately, these conventions are well proven and accepted and do not vary much between disciplines.

TECHNICAL INFORMATION – STATICS

- Statics involves mathematical and graphical presentations.
- The graphical presentations are well developed and accepted.
- Expect lots of conventions; they help to keep a complex subject under control.

Vector diagrams

Vector diagrams are a core method of expressing information about statics. They are used in both two- and three-dimensional situations.

REMINDER: ABOUT VECTORS

- Force is a vector – it has magnitude and direction.
- Vectors can be algebraically manipulated (using mathematics).
- Vectors can be represented graphically by a line of given length and direction.
- Vectors can be added using the 'parallelogram rule'.

BUT REMEMBER

- Stress is not a vector because it does not obey the laws of vector algebra; it is a tensor, and has its own set of rules.

Figure 7.1 shows a reminder of how 'force' vectors are drawn.

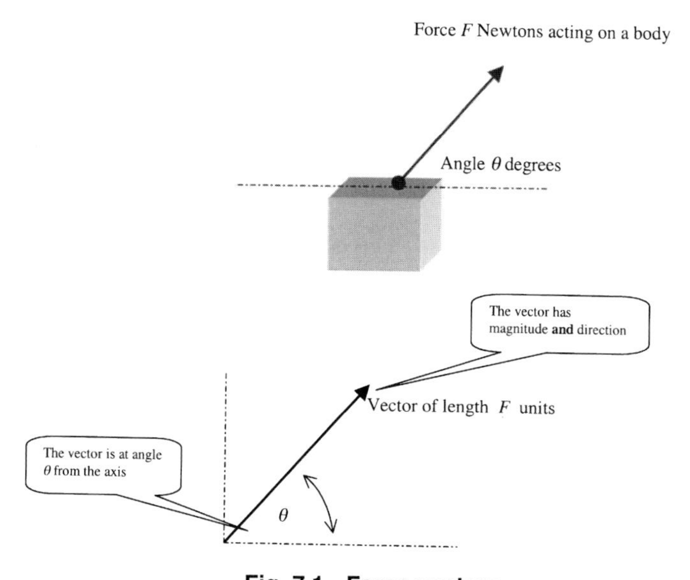

Fig. 7.1 Force vectors

Useful reference

http://web2.airmail.net/paz/FlightDisplay.html

Loading diagrams

There are many cases in statics where bodies and structures are acted upon by external loads. The conventional ways to portray these are given below.

Point loads

A point load may be caused actually by a point load, or it may be a representation of the weight of an object. Figure 7.2(a) shows the way that the weight of an object is approximated to a single 'point load' force acting through its centre of gravity.

Distributed loads

Distributed loadings occur when a load is distributed over the length or surface of an object. The distribution may be uniform or non-uniform and is conventionally shown by a series of arrows (see Fig. 7.2(b)). Note the conventions given below.

- The uniformly distributed load (UDL) is represented by a series of equally spaced arrows of the same length, or a wave symbol. The specific loading (N/m or N/m^2) is given algebraically.
- The non-uniform load is depicted by arrows of varying length, the longer arrows corresponding to the regions of greater loading. Note that the length of each arrow is indicative only, not a quantitative measure of the actual loading at that location. The numerical value of the loading at any point has to be determined by calculus or computerized techniques such as finite element analysis.

Forces on 'an element'

Many techniques in statics involve the consideration of forces acting on an element as a precursor to calculus techniques. This is common for stress analysis in shells, beams, tanks, tubes, and almost all real-world structures. Figure 7.2(c) shows how this is represented. Note that it can apply in one, two, or three dimensions. We will also see in Chapter 8 how a similar representation is used to show forces in dynamic bodies.

(a) Point loading assumed to act through the CG of an object

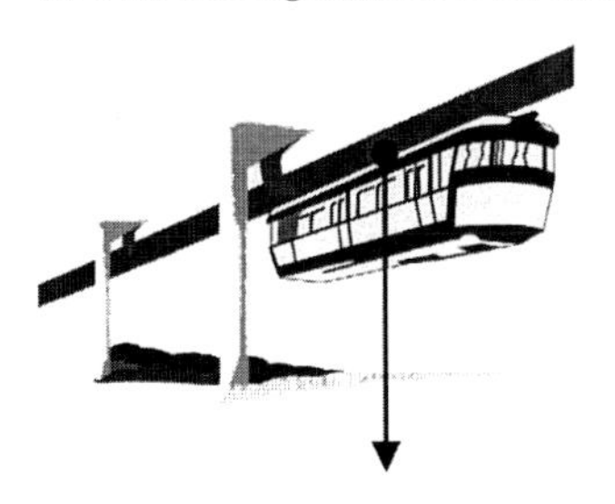

(b) Distributed loads

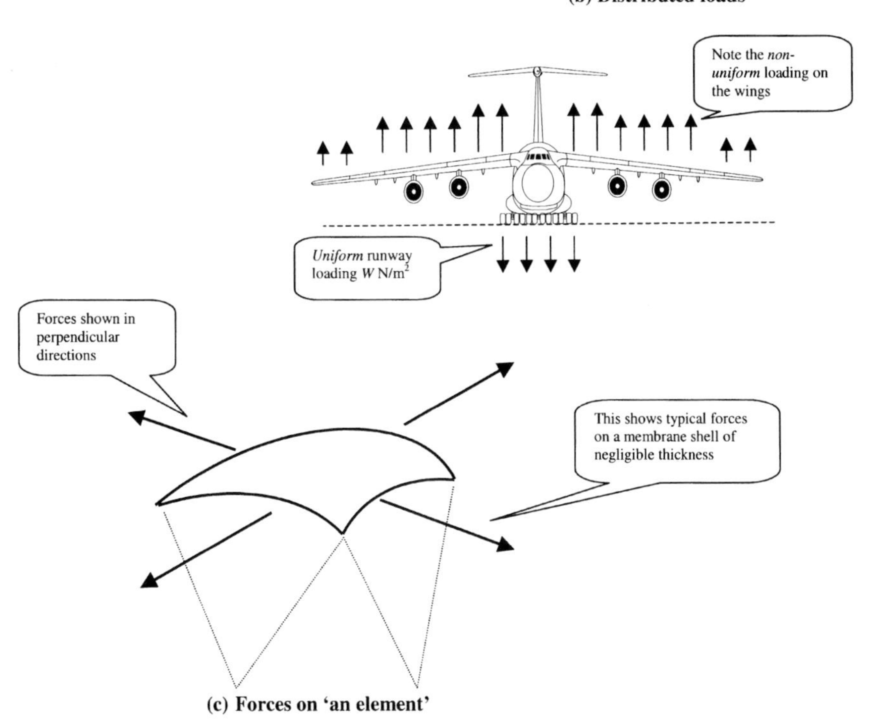

(c) Forces on 'an element'

Fig. 7.2 Point and distributed loadings

Framework diagrams

Analysis of engineering structures such as bridges is carried out using framework diagrams, as shown in Fig. 7.3.

FRAMEWORK DIAGRAMS – KEY FEATURES

- All structural members are represented by simple beams, pin-jointed at their ends, which are in either pure tension or pure compression.
- The triangular framework is 'simply supported', again by pin-joints, that give point load reaction forces.
- The inference is that the whole framework is in static equilibrium – it is not vibrating, wobbling, or moving about.
- Each framework member is identified by letters relating to the spaces on either side of it (Bow's notation).
- Forces in the members are found by constructing force polygons. There is an alternative way known as the 'method of sections'.

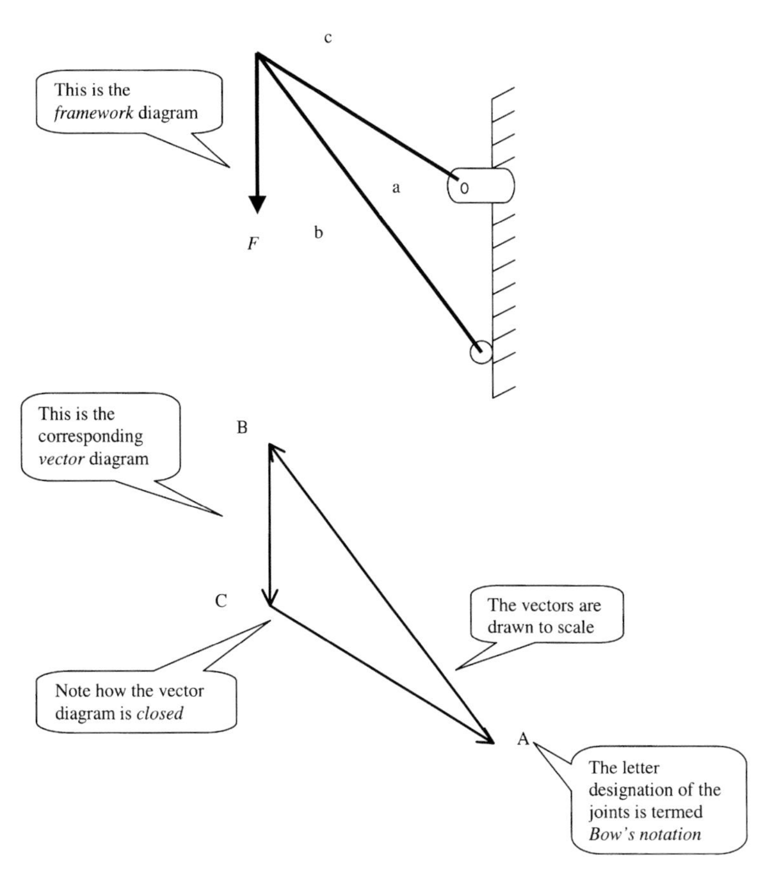

Fig. 7.3 Framework diagrams

Free body diagrams

Free body diagrams are a technique used to solve problems in statics and also some aspects of dynamics. Their purpose is to show every force acting on a body (or part of it) so that they can be resolved, and the values determined.

FREE BODY DIAGRAMS – THE PRINCIPLES

- A body (or part of it) is isolated by means of an imaginary system boundary.
- The isolated part is then considered as being in equilibrium.
- All the forces acting on the isolated part are shown in a diagram.

Figure 7.4 shows a typical example for a simple tower-crane structure. Note how the part that is isolated (in this case all of the crane) is shown in equilibrium 'while floating free in space'; it is this that enables the forces and moments to be calculated.

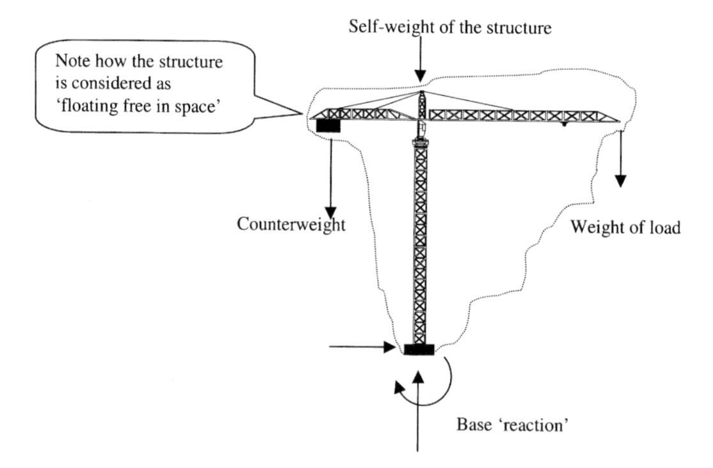

Fig. 7.4 A 'free body' diagram

Force transmission diagrams (structures)

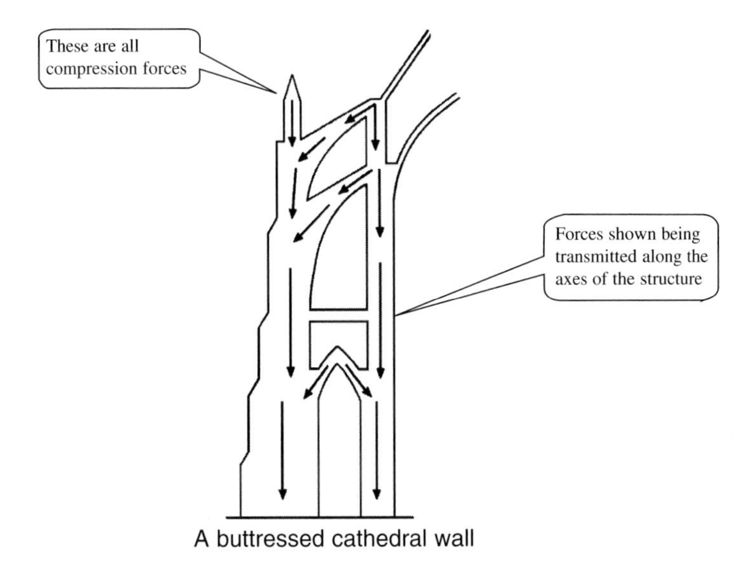

Fig. 7.5 A force transmission diagram

Structures are broadly defined as bodies that can resist applied forces while suffering only small deformations due to the elasticity of their materials of construction. This means that structures have forces acting within their components, which manifest themselves in several forms: tension, compression, bending, torsion, etc. There are a number of conventional presentation methods to represent these forces. Figure 7.5 shows one example for transmission of forces through a building structure.

FORCE TRANSMISSION DIAGRAMS – KEY FEATURES

- The transmission of forces is visualized as being along the axes of the various parts of the structure.
- Single-headed separate arrows are used to denote forces within the structure.
- The length of the arrows is indicative of the magnitude of the forces (compression forces in this example).

This type of simplified presentation is in common use in civil engineering and basic mechanical design.

A more complex example – 'load–line sketches'

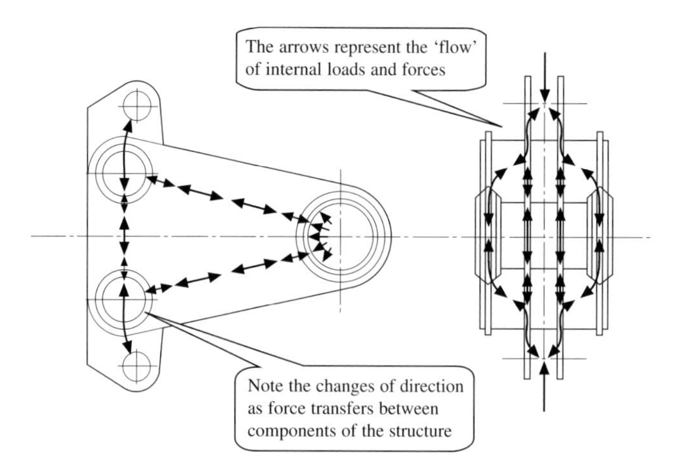

Fig. 7.6 A 'load–line' sketch

Figure 7.6 shows how internal forces are represented in a more complex component: an assembly of plates, tubes, and bosses. Such views are often termed 'load–line sketches'. Shown in several orthographic views, they are useful for showing how loads change direction in a structure, or for identifying parts of the structure which are redundant, i.e. which do not carry any load.

Tension diagrams

Structures which comprise a lot of ropes, wires, or thin structural members in tension are often represented by a dedicated tension diagram. Figure 7.7 shows an example for a crane derrick.

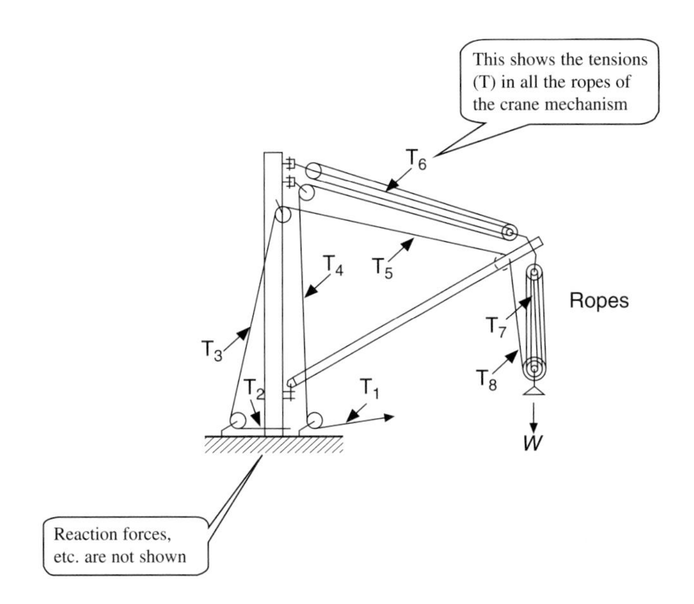

Fig. 7.7 A tension diagram

Detailed force diagrams

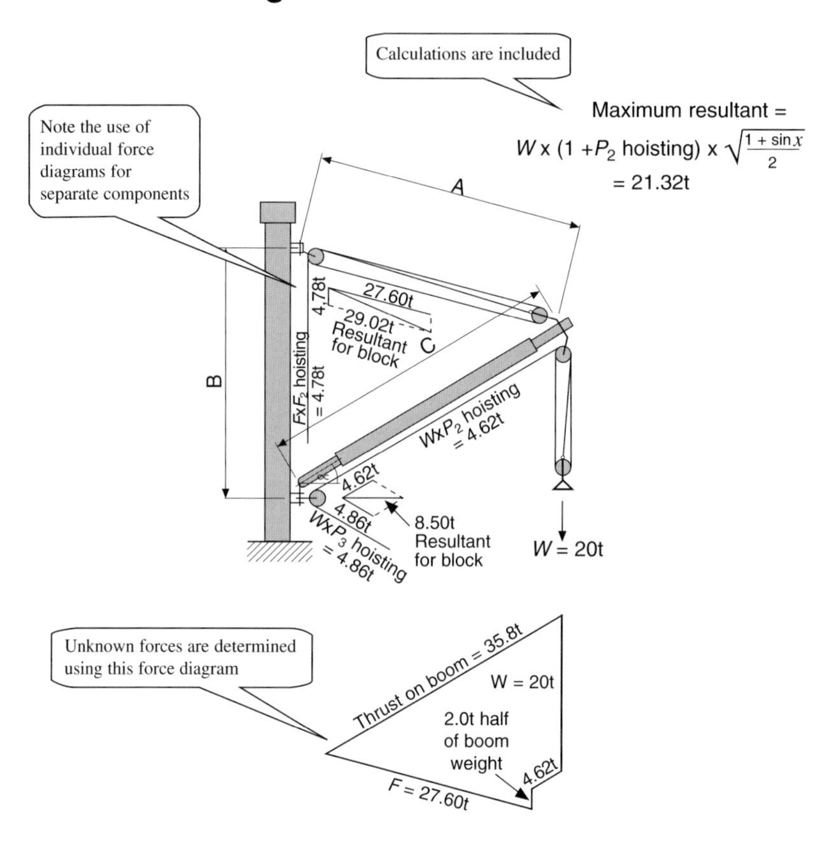

Fig. 7.8 A detailed force diagram

Detailed force diagrams are used for most engineering components at the early stages of the design phase. They are an amalgamation of several of the earlier presentation methods in this chapter (loading diagrams, framework analysis, etc.). Figure 7.8 shows an example for the crane derrick shown in Fig. 7.7. Note how the analysis results in a force diagram, allowing unknown forces to be calculated from the given information. Note also how aspects of dynamics such as friction forces are taken into account.

Forces in individual components

Forces in individual components of complex structures need to be analysed for the purpose of deciding the mechanical design of those components. This is done by combining the techniques of the free body diagram and the vector resolution of forces as described earlier. Figure 7.9 shows the application to two components of the crane derrick.

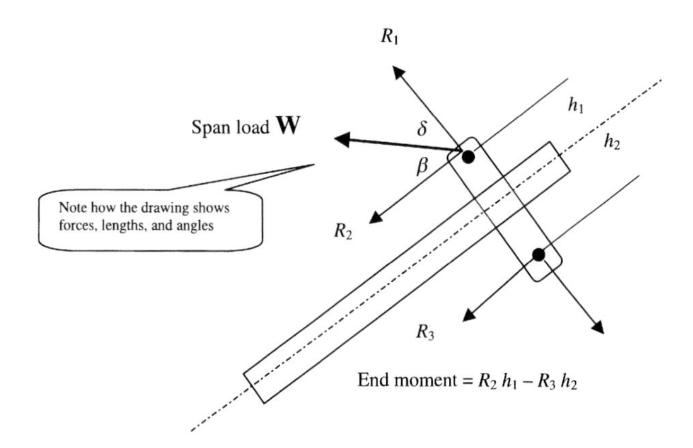

Fig. 7.9 Diagram showing forces in individual components

Shear force/bending moment diagrams

Shear force (SF) and bending moment (BM) diagrams are used to analyse structural components (members) which are subject to bending. Shear force is an internal force acting to prevent parts of the member sliding relative to each other. Bending moment is the corresponding internal moment resisting the couples produced by the externally applied loads and moments. SF/BM diagrams are presented as shown in Fig. 7.10.

SF/BM DIAGRAMS – KEY FEATURES

- The member, SF, and BM diagrams are shown vertically above each other, as a set.
- 'Sagging' bending moments are classed as positive (+M) and 'hogging' moments as negative (-M).
- The +M/-M sign convention is decided by the curvature of the member rather than the direction of its deflection up or down, i.e.

  -M is convex upwards shape.

  +M is concave upwards shape.

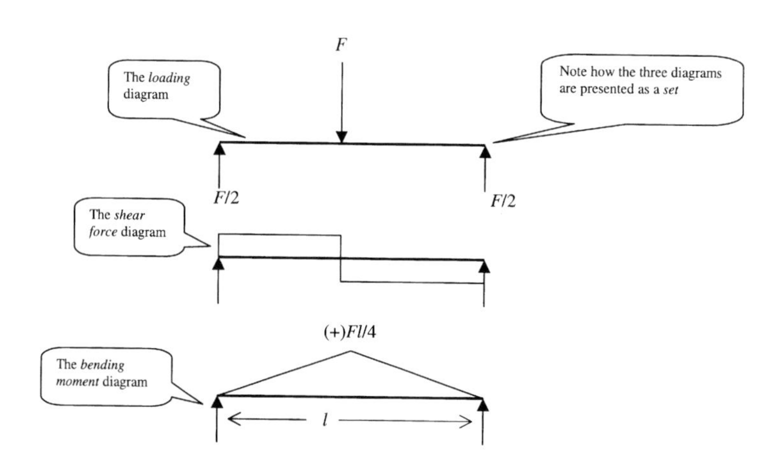

Fig. 7.10 Shear force and bending moment (SF/BM) diagrams

Useful references

http://members.aol.com/ADBell/sfbm.htm

http://www.npiec.on.ca/~echoscan/43-16.htm

Multiaxial stress diagrams

Multiaxial stresses are a real feature of the engineering world; most static or dynamic engineering components are subject to multiaxial ('complex') stresses in use. Multiaxial stresses are not easy to represent so the convention is to consider stress 'on an element'. A small 2-D slice or 3-D element of the component is shown, with the stresses represented as single arrows ('tensors'). Remember that these stress arrows are not vectors and do not obey the laws of vector algebra. Figure 7.11 shows an example, in this case relating to a three-dimensional stress state. Note the features:

- the idea of an infinitesimal element taken from within the body of a physical component;
- the representation of stress by single tensor arrows.

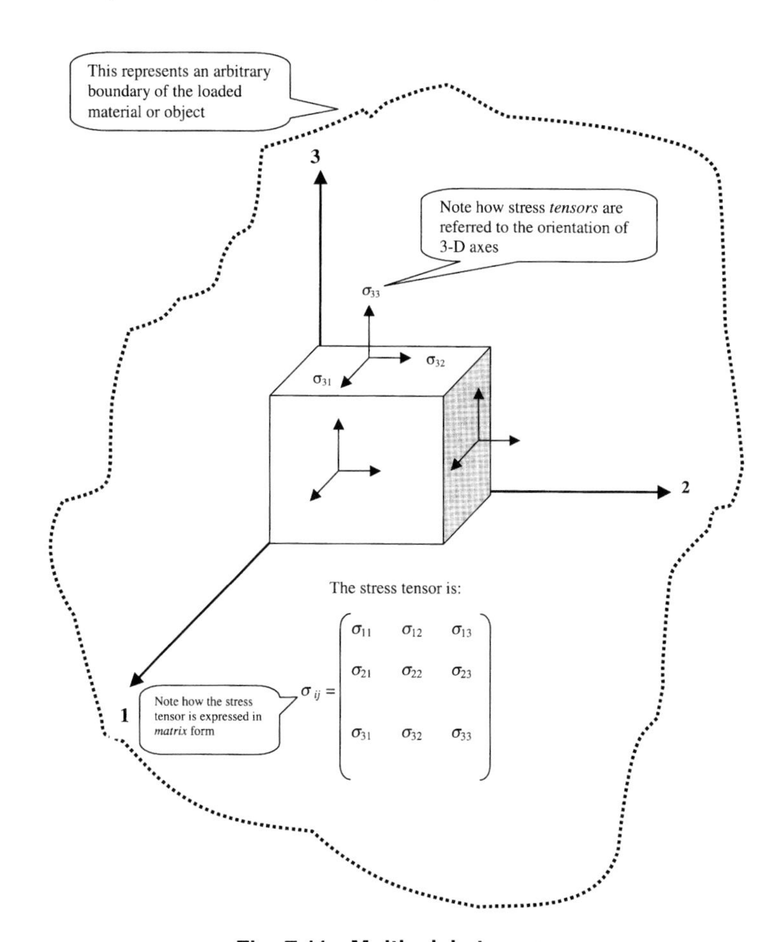

Fig. 7.11 Multiaxial stresses

Torsion diagrams

Twisting or *torsion* of engineering components such as shafts, bolts, discs, etc. is a common feature of engineering design and is represented in a number of conventional ways. Figure 7.12(a) shows the conventional use of 'twist arrows', indicating the sense in which an object is being twisted.

Figure 7.12(b) shows the way that the shear stress (τ), induced by the torsion, is represented. The value varies across the diameter of the component, hence the representation by different lengths of tensor (shear stress, like tensile stress, is a tensor quantity).

Figure 7.12(c) shows how the effects of the shear stresses are represented, i.e. it results in complementary stresses at 90° to the plane of shear stress. This is conventionally shown as a square element of undefined size. Remember that diagams like this showing 'elements' of a body are conceptual, rather than strictly physical.

(a) Showing torsion using simple 'twist arrows'

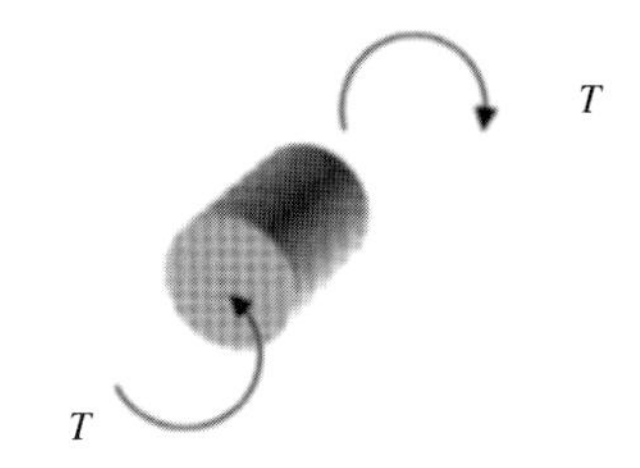

(b) Representing torsion by the shear stress (τ) across the section

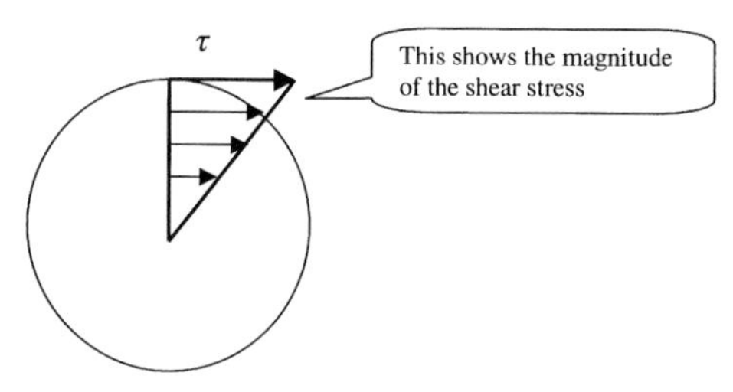

(c) Showing the *effects* of torsion, i.e. complementary stresses

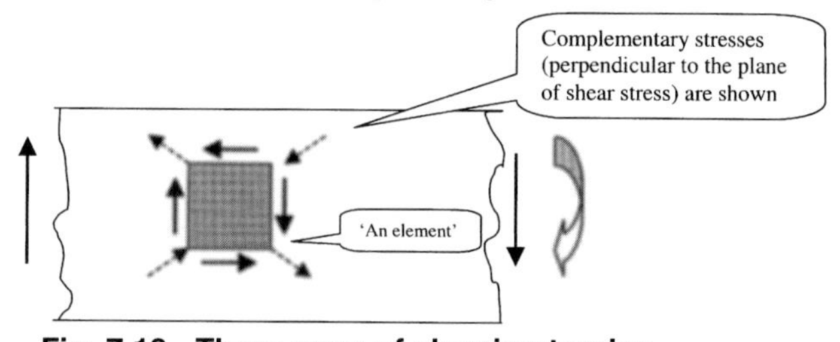

Fig. 7.12 Three ways of showing torsion

Deflection and distortion sketches

No materials are absolutely rigid, so the amount by which a component deflects or distorts elastically in use is a key consideration in engineering design. Complex assemblies of rotating or mating parts in all types of machines and pressure vessels are analysed in depth at the design stage to investigate how the components change shape in use. The main difficulty in showing deflections on a sketch or computer simulation is that the amount of deflected movement is generally small, relative to the size of the object that is being deflected. This means that in order to show the deflections, they need to be exaggerated. This is common practice in all the mechanical disciplines. Figure 7.13 shows the usual method for a mechanical component.

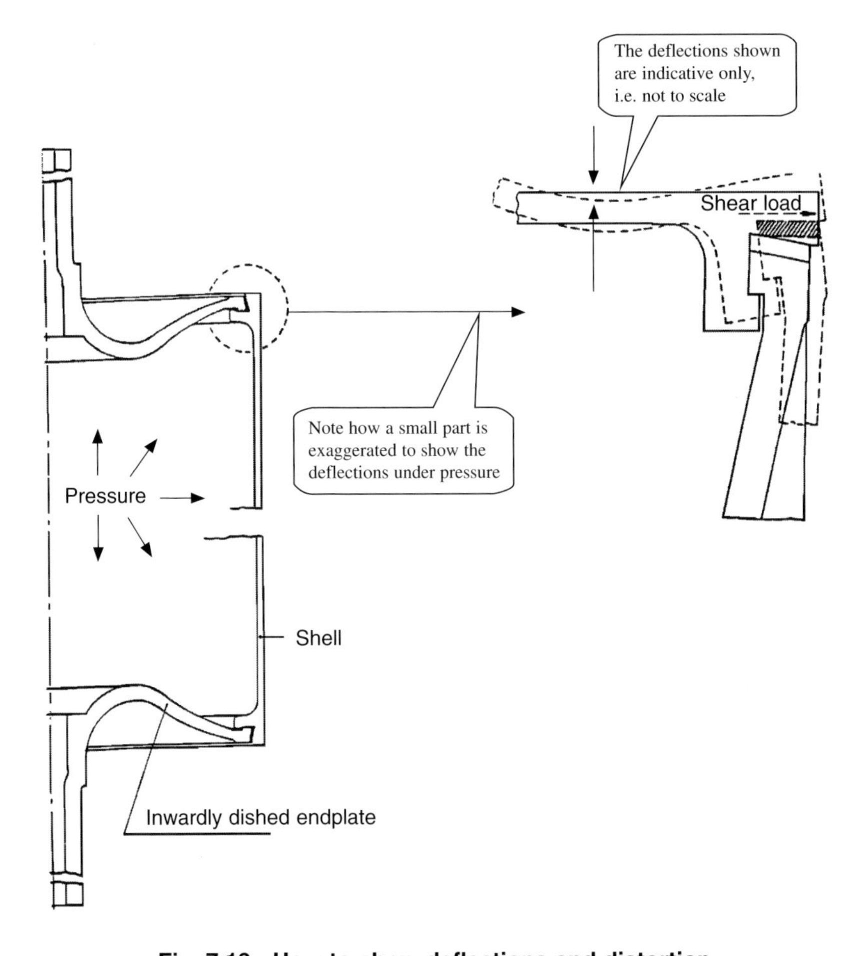

Fig. 7.13 How to show deflections and distortion

FE model plots

Distortions can also be shown by displaying a computer-generated finite element (FE) 'plot' of a component before and after distortion. Figure 7.14 shows an example.

FE 'DISTORTION PLOTS' – KEY FEATURES

- They are best restricted to 2-D displays, otherwise they can become very complicated.
- The deformed part is shown at an exaggerated scale (up to 25 or 30 times larger) to clearly show the extent of the distortion.
- The FE 'mesh' is normally shown, to give some indication of the nature and intensity of the stresses that are causing the distortion.

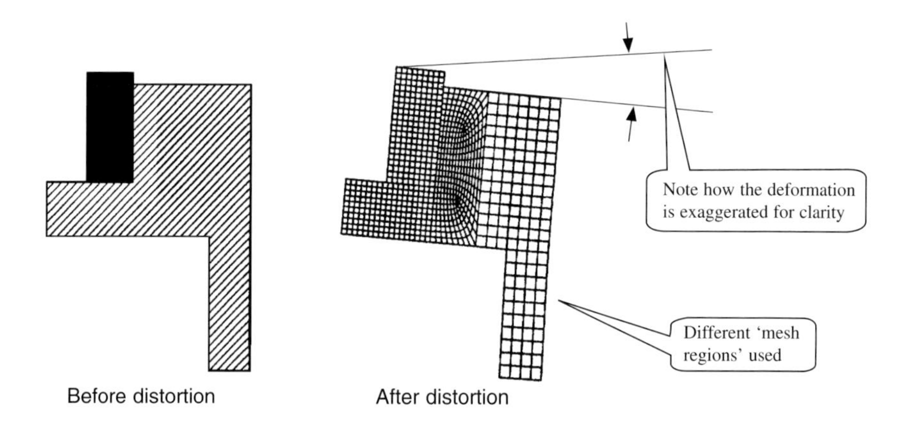

Fig. 7.14 A FE 'distortion' plot

Misalignment diagrams

Misalignment of mechanical components is an important criterion, because it can result in undesirable high forces and moments. For this reason, complex designs often have a misalignment analysis carried out to show the extent, and consequences, of misalignments that can be anticipated. Two types of diagram are used: simplified 2-D and 3-D. Note two key features illustrated in the typical example for a short coupling (see Fig. 7.15).

- The misalignment is exaggerated for clarity (to a scale of up to 25–30 times larger in some cases).
- The forces and moments resulting from the misalignment are indicated.

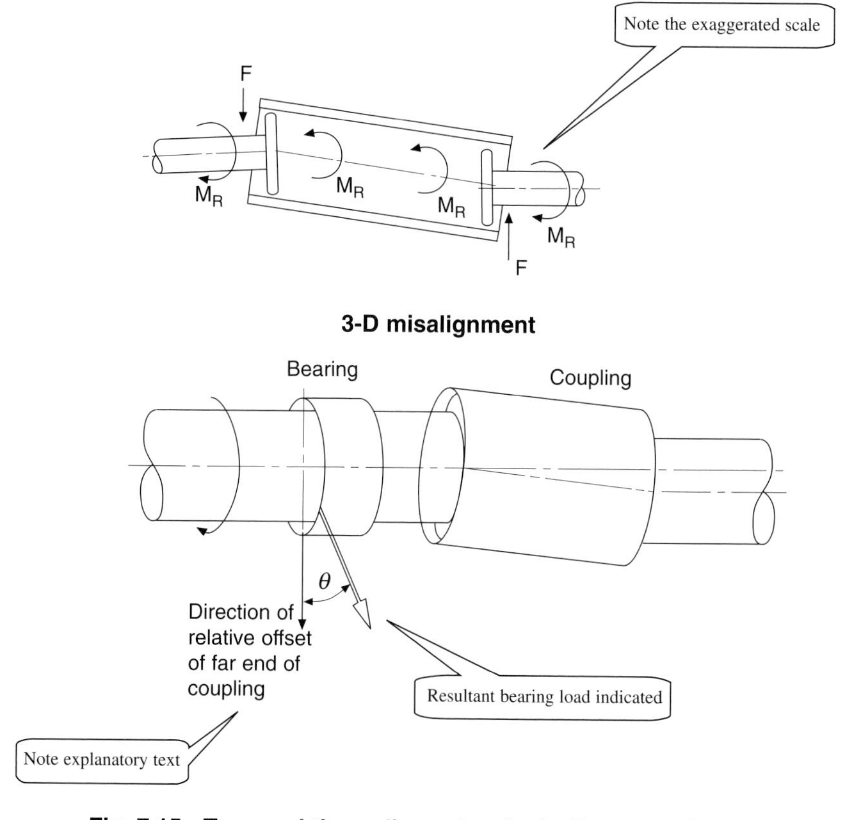

Fig. 7.15 Two- and three-dimensional misalignment diagrams

Chapter 8

Mechanics – Dynamics

Introduction to dynamics – what is it?

Dynamics is the branch of mechanics dealing with things that move. It is divided broadly into kinematics (the motion of mechanical systems) and kinetics (the forces related to motion). Similar to statics (covered in Chapter 7), dynamics is essentially a design discipline.

Dynamics – its technical information

Showing how things move is an important aspect of presenting technical information. It is necessary in several technical disciplines to communicate, on a printed page, details of how physical components move. There are four main ways to do this:

FOUR WAYS TO REPRESENT MOTION

- Using mathematics, i.e. equations of motion.
- Using vectors.
- Showing accurate geometrical 'paths' of the motion.
- Purely pictorial representations.

It is worth summarizing these four techniques:

Showing motion using mathematics
This is the most prescriptive way. The existence of Newton's laws of motion (which apply to everything that moves) means that all movement can be described in algebraic terms. This is accurate and theoretically robust, but will not be understood by all readers – particularly the non-technical ones. Algebraic expressions are, therefore, often used to support other methods of depicting motion.

Showing motion using vectors
Vectors representing motion have their roots in the study of kinematics. Vectors show the magnitude and direction of motion, can be made to represent velocity or acceleration, and are easily adapted to depict linear motion or motion in a circle or ellipse. Again, their disadvantage is that non-technical people may not understand them.

Showing motion using geometrical 'paths'
This includes graphical presentations such as loci (see Chapter 3) in which the path of a moving object or point is drawn in an accurate way either by computer or conventional drafting techniques. The motion may then be additionally described using algebraic expressions. This result is easy to understand, but can be difficult and/or expensive to produce.

Showing motion pictorially
Pictorial representations portray the motion of an object or point but do not give an accurate description of that motion. They provide a visual impression only and can be shown in 2-D or 3-D, depending on the application. Pictorial views can be assimilated easily by technical or non-technical readers.

Distance/velocity/acceleration diagrams

These are a set of diagrams used mainly to depict linear motion. At their root are the basic relationships between distance (s), velocity (v), and acceleration (a) These can be expressed in normal algebraic or calculus terms.

> **DISTANCE/VELOCITY/ACCELERATION DIAGRAMS – KEY FEATURES**
>
> - The diagrams are displayed as a set, set out in a vertical format in the order shown in Fig. 8.1.
> - All three diagrams have time as the horizontal axis; it is only the vertical axis that varies.
> - Because s, v, and a are linked by calculus, the areas under the curves have significance. They are sometimes shaded or hatched to reflect this.
> - Acceleration is the most important parameter because it results in mechanical stresses, which influences the mechanical design of a component.

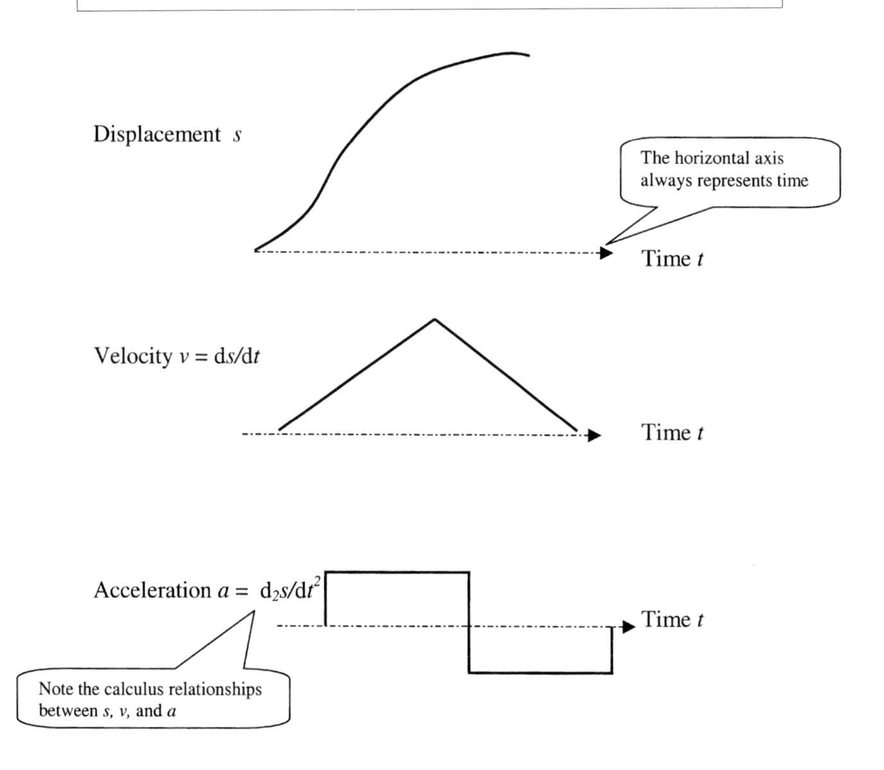

Fig. 8.1 Distance/velocity/acceleration diagrams

Relative motion diagrams

In most practical technical disciplines, it is rare to have to describe the motion of a point or component relative to a set of axes that are fixed absolutely in space. More commonly, motion has to be described in relation to another moving set of axes, or part of a body. This is known as 'relative motion'. The conventional way to show this, and to solve motion problems, is by using vectors. Figure 8.2 shows a typical presentation of relative motion.

RELATIVE MOTION DIAGRAMS – KEY FEATURES

- The diagram showing the motion is not to scale, it is a sketch only.
- Motion is shown by single arrows, with magnitude and direction.
- The accompanying vector diagram has to be to scale.

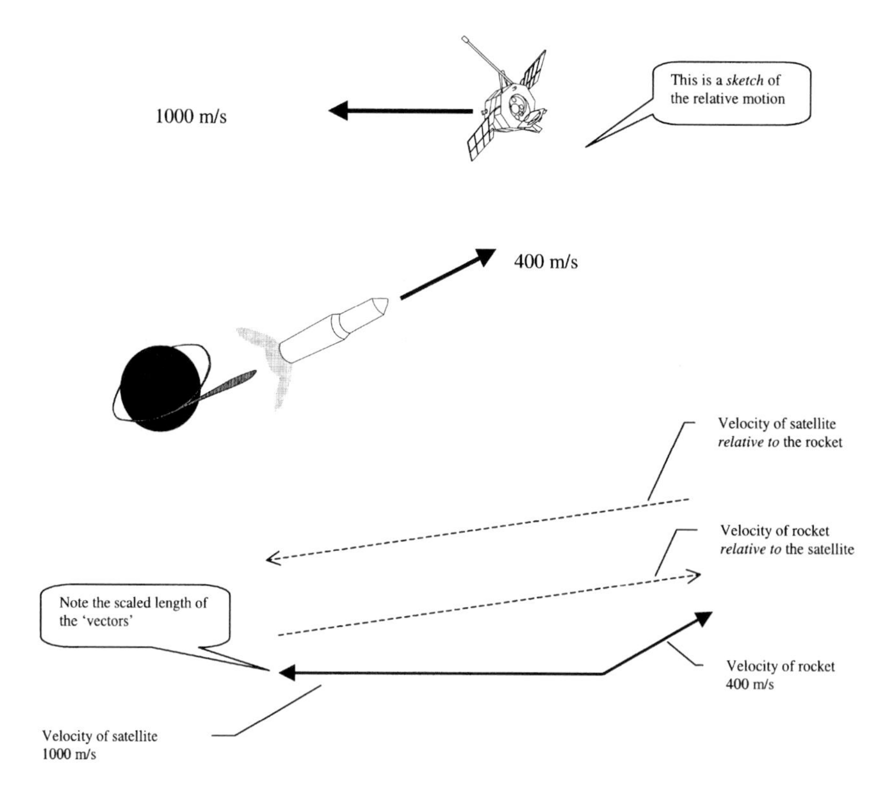

Fig. 8.2 The concept of relative motion

Irregular motion diagrams

There are many examples of irregular motion in the world, particularly in the applied technical disciplines where information may need to be presented in a less theoretically rigorous way. Irregular motion is difficult to describe using simple vectors or mathematics, so it is best shown using graphical or pictorial methods, depending on how complex the motion is.

Simple movements

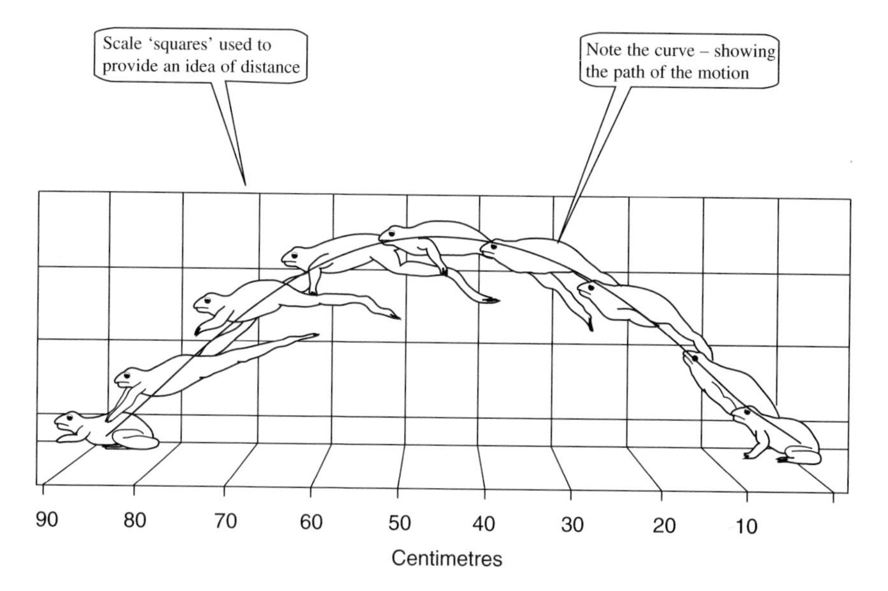

Fig. 8.3 A simple 'graphical' motion diagram

Figure 8.3 shows a simple case – the motion of a leaping frog. The frog is making a single, unimpeded leap, which follows a certain geometric path (defined by the laws of physics), so it does contain some regularity. The motion is shown graphically. To describe the motion fully, however, would require multiple vector diagrams.

SIMPLE 'GRAPHICAL' MOTION DIAGRAMS – KEY FEATURES

- A broad horizontal and vertical scale.
- The use of simple illustrations to liven up the presentation.
- A curve drawn along the line of motion.

Complex movements

Some motions are so highly irregular that to depict them accurately is almost impossible without photographic or computer equipment. In such cases, the only practical way to depict the motion is in an indicative way, using pictorial presentations. Figure 8.4 shows an example from the natural world, i.e. two separate types of movement of a snake. In Fig. 8.4(a) the movement is restricted to a type of wave motion, but this is still so complex that a reasonably accurate graphical representation, as in Fig. 8.3, would be impossible. Note how a set of three pictures is used to give an impression of the motion. A reader will feel that they can at least picture the motion in their own mind, even if they could not go on to draw an accurate motion diagram, or describe it by vectors or mathematics.

Figure 8.4(b) shows an even more complex example, where the motion is totally unconstrained in two dimensions. Notice how it is now more difficult to depict the movement with the same three pictures. Figure 8.4(c) is an example of unconstrained and irregular motion in three dimensions. It is next to impossible to portray this accurately on a printed page, or even by mathematics.

(a) Constrained – snake in a tube

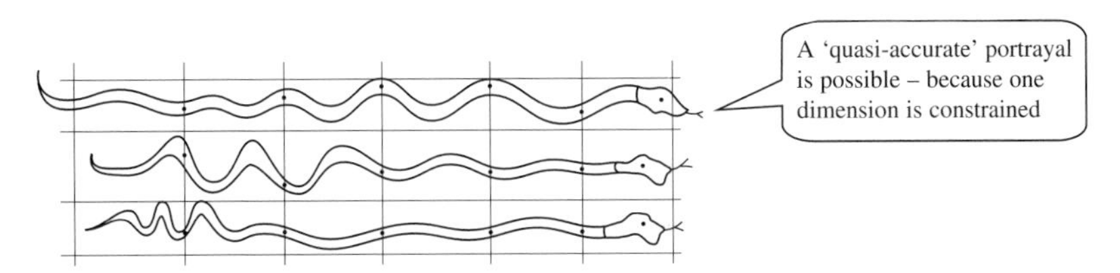

(b) Unconstrained snake motion – the 'sidewinder'

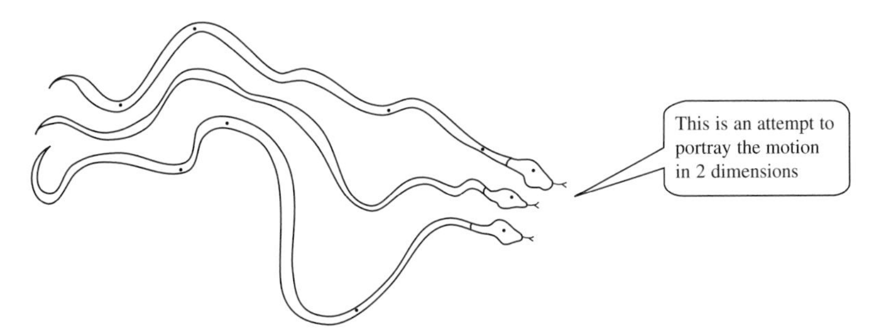

(c) Complex and unrestrained 3-D movement

Fig. 8.4 Complex movements

Mechanism diagrams

Engineering mechanisms come in many shapes and sizes, from simple two-element linkages to complex and interlinked mechanical systems. The motion of mechanisms is depicted by using geometrical paths. This requires an accurate development of the paths or loci of the moving parts, sometimes termed 'projections'. Figure 8.5 shows a typical example, in this case for a cam mechanism.

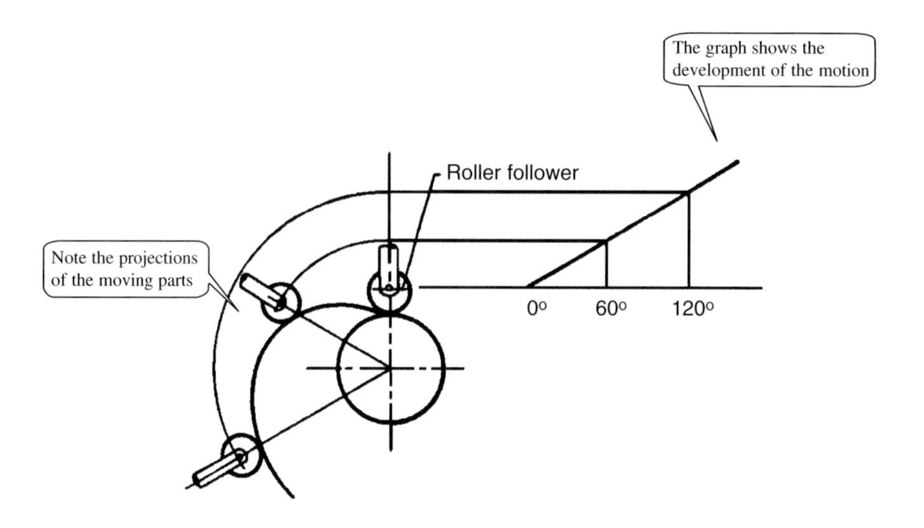

Fig. 8.5 A mechanism diagram

MECHANISM DIAGRAMS – KEY FEATURES

- The projection of the pivot point of the cam as it translates (moves in a curve).
- The illustration of the cam follower in three separate locations during its movement. This helps with visualization of the movement of the mechanism.
- Simplified but accurate drawing, uncluttered by irrelevant technical detail.

Displacement diagrams

The motion of mechanisms such as cams and complex linkages is sometimes best described accurately using a displacement diagram. This displays the linear or angular displacement of a particular point on the mechanism with respect to a parameter of the motion such as angular rotation or time. Again, this is an accurate, graphical way of displaying motion. Figure 8.6 shows an example.

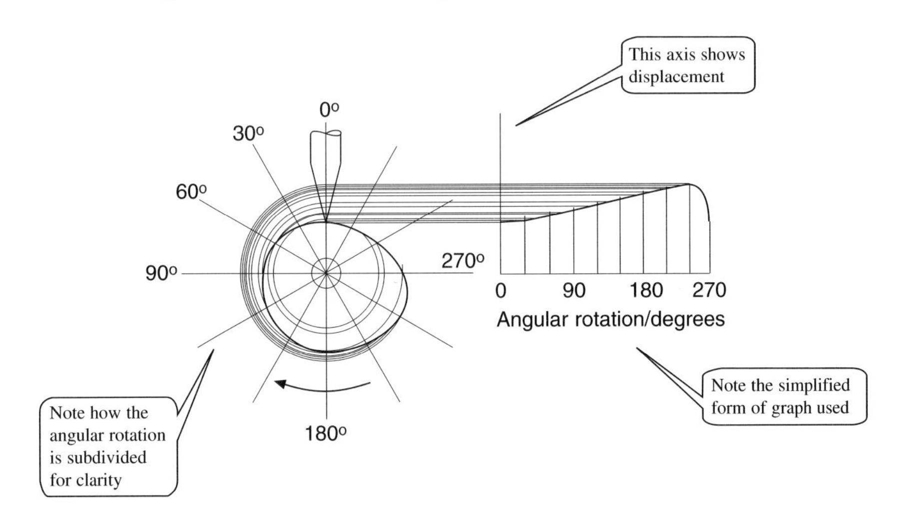

Fig. 8.6 A mechanism 'displacement' diagram

'Loose mechanism' motion diagrams

'Loose mechanisms' is a generic term applied to loosely specified mechanism linkages whose purpose is to show an approximate form of movement. Compare this with the previous types that are closely specified to produce a well-defined, accurate form of motion. Loose mechanism motion diagrams are used to depict:

- the motion of live forms such as humans and animals;
- physical constructions with large movements and deflections such as bridges, roads, and some types of building;
- loosely interrelated mechanical systems of any type, particularly those with non-linear, irregular motions.

Figure 8.7 show a typical application in the natural world.

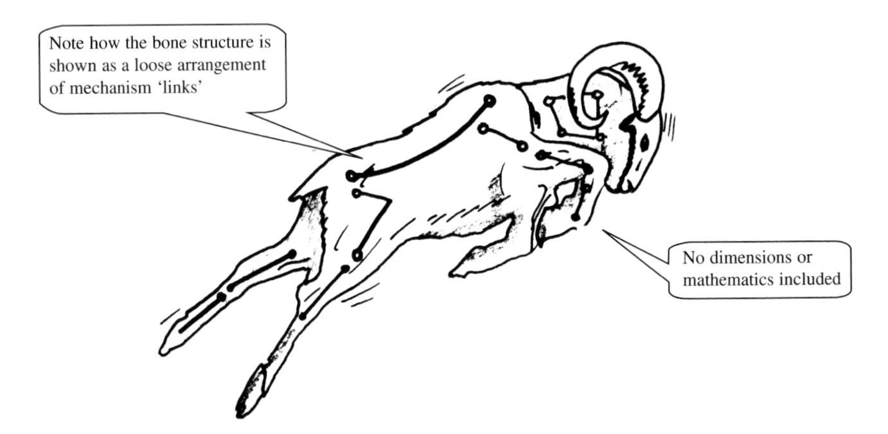

Fig. 8.7 A 'loose mechanism' motion diagram

Simple harmonic motion diagrams

Simple harmonic motion (shm) is a specific type of motion in which the displacement, velocity, and acceleration of a point are sinusoidal functions of time. The assumption of shm is commonly used when discussing subjects such as circular motion and vibration. The conventional way of displaying shm is by a combination of graphical sketch and mathematical equations. The equations describe the motion rigorously and accurately, while the sketch is included to help visual interpretation (i.e. for non-mathematical readers). Figure 8.8 shows a generic example that can be used for any shm application.

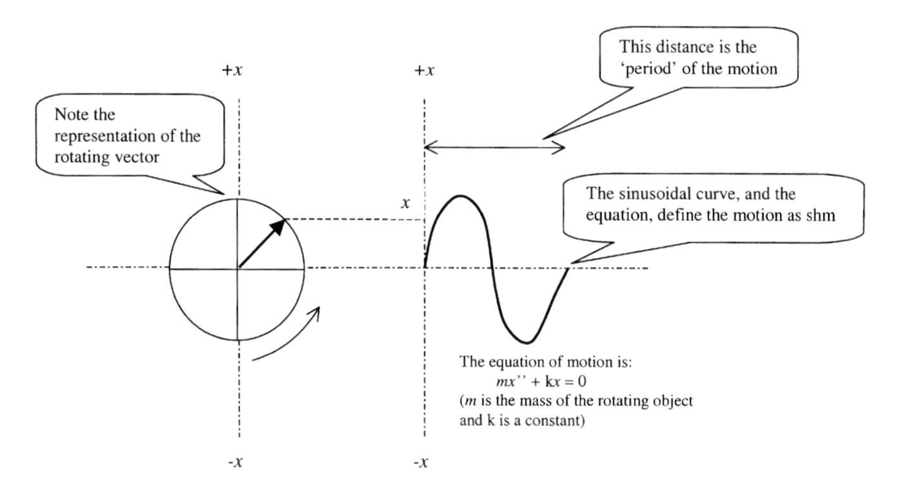

Fig. 8.8 Simple harmonic motion (shm)

Wave motion diagrams

Representations of wave motion are used in technical disciplines such as physics, acoustics, electronics, optics, dynamics, and quantum theory. Both longitudinal and transverse types of waves can be portrayed by the simple wave diagrams shown in Fig. 8.9. Most wave diagrams are not drawn to scale, i.e. they are a pictorial representation only, so need the addition of technical data such as 'legends', to describe the motion accurately.

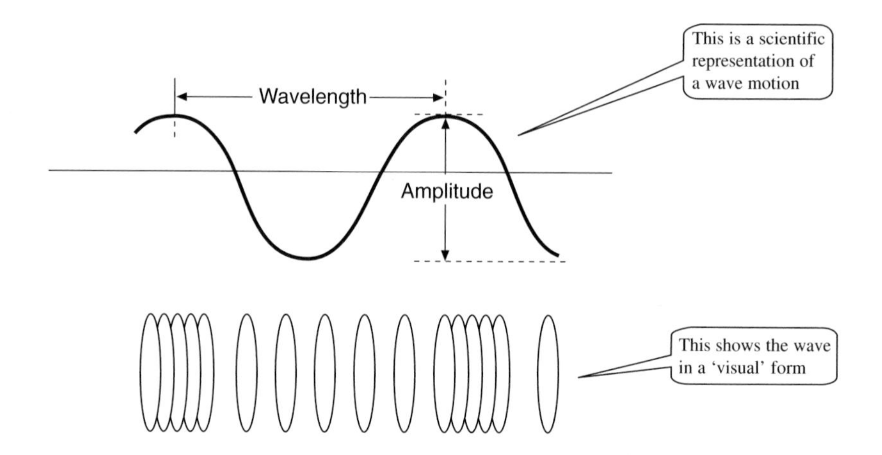

Fig. 8.9 Two ways to show wave motion

And then: strange motion – the tractrix

There is nothing very unusual about the tractrix. It has important connotations for the definitions of Euclidean and Lobachevsky geometry (if you need to know), but apart from that it is just an interesting type of motion path. Figure 8.10 shows the type of motion that generates a tractrix, and a purely geometrical method that gives the same result (if you need to draw one).

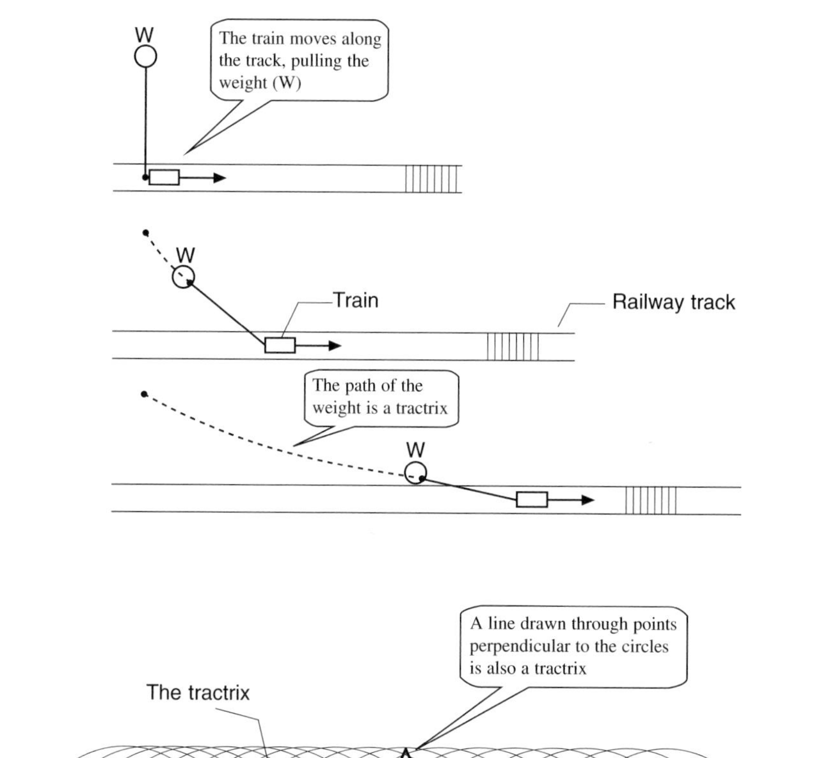

Fig. 8.10 Strange motion – the tractrix

Chapter 9

Project Management Information

Introduction – what information?

Project management, as a discipline, involves a great variety of types of technical information. It comprises not only the technical information relevant to a project (mechanical, civil, research, software engineering, or whatever) but also a complete set of supporting information relating to the techniques of project management, such as project planning, scheduling, and performance assessment.

Projects do not have to be large and complex to require the correct presentation of project management information. Small, simple projects are used all the time in the technical world. Any programme of work which is undertaken to achieve a specific objective can be defined as a project, and all have management information attached to them.

To complicate matters, project management information varies in form from heavily conceptual information, about strategy and development for instance, to hard, quantitative data about timescales, costs, and revenues. This is balanced by the fact that ways of presenting the information tend to be heavily standardized. There are well-accepted technical standards in Europe, USA, and Asia that are surprisingly similar.

PROJECT MANAGEMENT INFORMATION – KEY POINTS

- Ways of presenting project management information are much the same for large or small projects.
- Presentation methods vary from conceptual to quantitative. Expect to see a bit of everything.
- The methods of presentation are more or less standardized and underpinned by published technical standards.

Project life cycle diagrams

All projects have a life cycle. It takes the form of a sequence of (sometimes overlapping) phases. Although the number of phases may vary with project size and complexity, the chronological order in which they occur generally does not. Figure 9.1 shows a common way that the life cycle is depicted; in this example it contains five phases.

PROJECT LIFE CYCLE DIAGRAMS – KEY FEATURES

- The diagram is a conceptual representation, and does not pretend to be a project timetable.
- Major project milestones are included. These are statements of 'intent' rather than firm date commitments.
- Note the simplicity of the diagram – there is no need for elaborate graphics or shapes. It shows the basic cycle phases only, and makes no attempt to convey any further information about the character of the project.

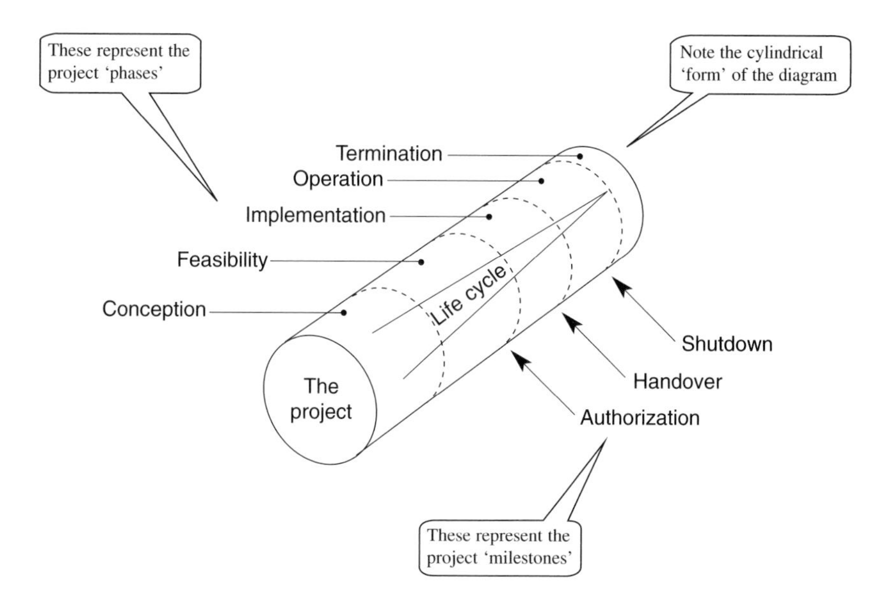

Fig. 9.1 A basic project life cycle diagram

Useful reference

BS 6079: 1996 *Guide to Project Management*. The British Standards Institution.

Construction programme diagrams

The construction programme forms the 'global view' of a project. It is conventionally shown as a bar chart representation of the project, with all the main tasks shown on the horizontal axis.

CONSTRUCTION PROGRAMME DIAGRAMS CONTAIN THREE PARTS

- A histogram and cumulative curve. This shows the allocation of resources (manpower, etc.) in both histogram and cumulative curve form. The histogram bars give an indication of weekly or monthly resource requirements, while the cumulative curve summarizes information for costing purposes.
- A bar chart of project tasks. This shows the itemized tasks necessary to complete the project.
- A task network. This acts as a preliminary stage to the later compilation of a full critical path analysis (CPA) network for the project. It shows which activities have to occur before (or after) others, depending on the physical or procedural constraints of the project.

Figure 9.2 shows a typical example, in this case for a bridge construction project. Note the way that significant factual information is provided in the diagram, even though it is essentially a visual presentation technique.

Useful references

http://louisa.levels.unisa.edu.au/se1/
http://ngst.gsfc.nasa.gov/project/text/toptencharts.html

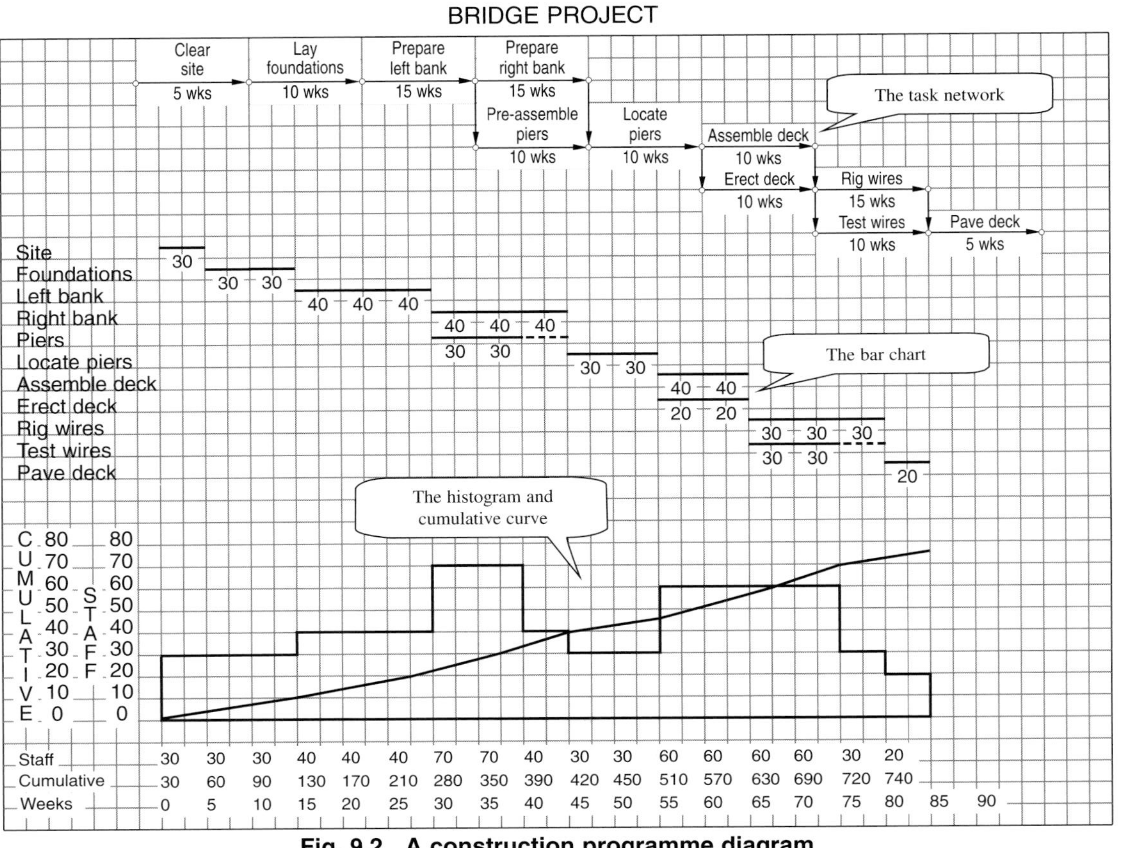

Fig. 9.2 A construction programme diagram

Activity sequence charts

Projects that have a lot of small and highly overlapping activities generally need a separate activity sequence chart. It shows the activities set against a (horizontal axis) timescale. The 'certain' period of each activity is shown as a shaded block with any possible extension shown in outline, or as an arrow (see Fig. 9.3). These charts are often entered manually into a project management software system where the information is further processed into critical path network format.

> **ACTIVITY SEQUENCE CHARTS – KEY FEATURES**
>
> - Large projects can have hundreds of activity entries (vertical axis).
> - Time is always shown on the horizontal axis.
> - Use simple block shading and lines only – no fancy graphics or textures.

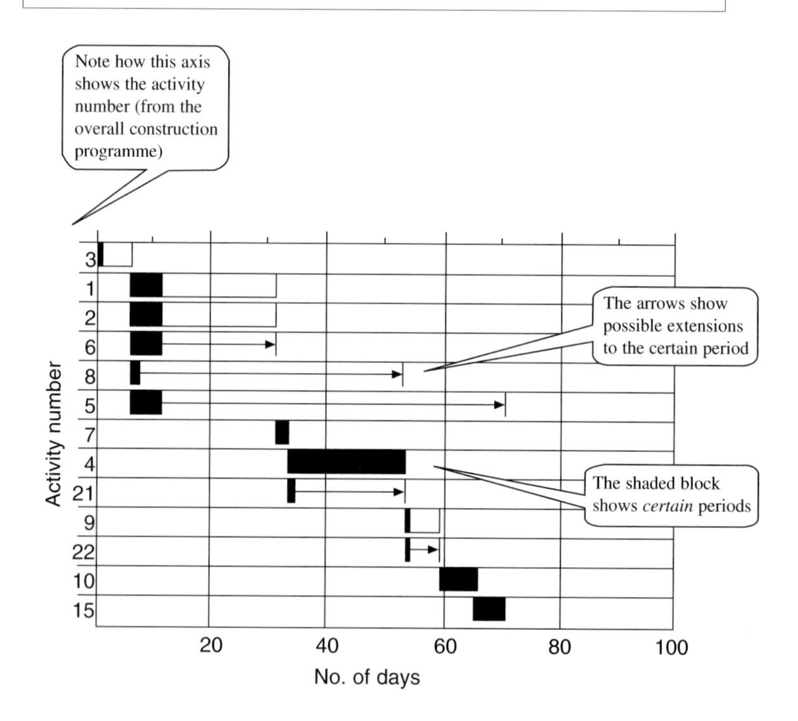

Fig. 9.3 An activity sequence chart

Network charts

These are also known as critical path analysis (CPA) charts or sometimes programme evaluation and review technique (PERT) charts. They are relevant to projects and programmes of all sizes and used extensively in software packages. The chart is based on the project timescale, showing not only the order in which project activities are done but also the duration of each activity. The charts are traditionally shown as a network of linked circles, each containing three pieces of information (see Fig. 9.4). The critical path is the path through the network that has zero float and it is shown as a thick, arrowed line. Float is defined as the amount of time that an activity can shift, without affecting the pattern or completion date of the project.

> **NETWORK (CPA) CHARTS – KEY FEATURES**
>
> - They are usually orientated horizontally; vertical ones are more difficult to assimilate.
> - Text labelling is included to describe the activities.
> - Large projects often have multi-sheet coloured charts; there can be hundreds of activities shown.

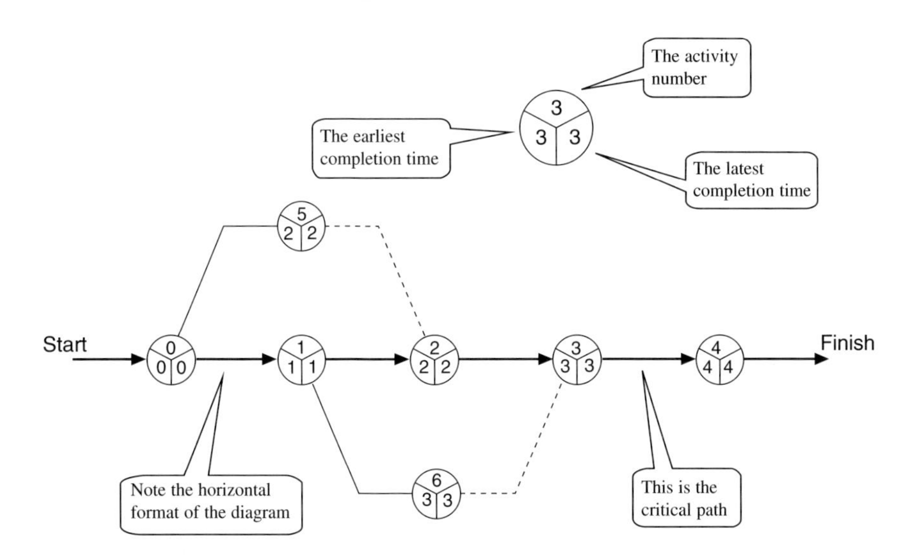

Fig. 9.4 A basic CPA 'network' chart

Project earned value (PEV) charts

The PEV chart is one of the mechanisms of project performance analysis. It involves assigning value to project work achieved and comparing this with the planned cost of reaching particular project milestones. The technique is sometimes referred to as budgeted cost of work performed (BCWP). This type of chart is a graphical replacement for older, tabular ways of comparing actual versus planned cost. Figure 9.5 shows a typical example for a construction project.

PEV CHARTS – KEY FEATURES

- Format. Note that the chart timescale (horizontal axis) extends over the full horizontal period of the project with the 'time now' clearly shown. Hence, the situation to the left of the 'time now' is actual, while that to the right is predicted.
- Cost axis. The cost axis is always vertical.
- Standard abbreviations. Figure 9.5 shows the conventional abbreviations used in the construction industry.
- The importance of variances. The whole point of the PEV chart is to show cost variances. These are often shown in colour, or as bold arrows and typeface.

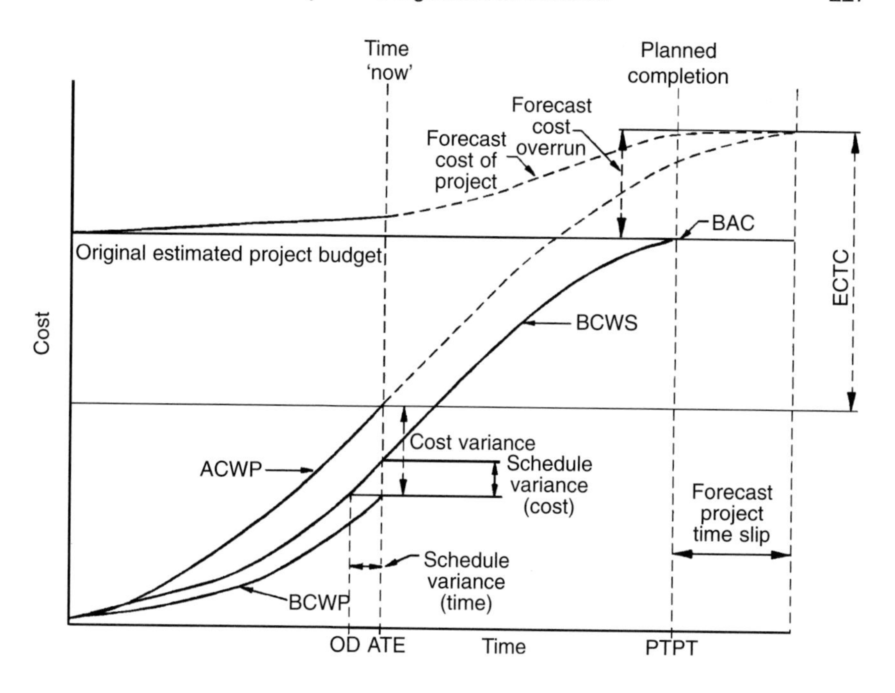

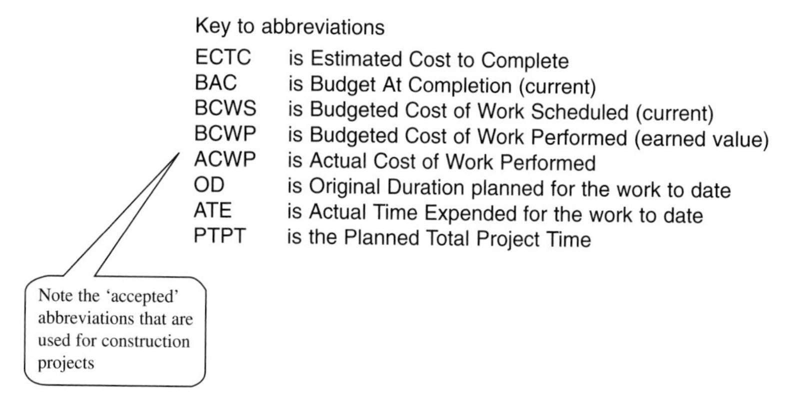
Key to abbreviations

ECTC	is Estimated Cost to Complete
BAC	is Budget At Completion (current)
BCWS	is Budgeted Cost of Work Scheduled (current)
BCWP	is Budgeted Cost of Work Performed (earned value)
ACWP	is Actual Cost of Work Performed
OD	is Original Duration planned for the work to date
ATE	is Actual Time Expended for the work to date
PTPT	is the Planned Total Project Time

Fig. 9.5 A project earned value (PEV) chart

‘Bulls-eye’ project diagrams

The formal name for these is ‘project cost funding estimate charts’. They are used to analyse the effect of recovery action initiated when a project has deviated too much from its cost or timescale targets (i.e. its variances are too high). The diagram is in graph form and shows the cumulative schedule variances (horizontal axis) plotted against cumulative cost variance (vertical axis). The progress of these variances over time is then shown by plot points on the chart. The intersection point of the axes (the ‘bulls-eye’ point) represents the state where the project has zero schedule and cost variance and is, therefore, ‘on track’. Note the key features in Fig. 9.6 as outlined below.

‘BULLS-EYE’ PROJECT DIAGRAMS – KEY FEATURES

- The plot points are joined by lines, showing the trend of the variances as the project timescale progresses.
- The chart is easily assimilated. The greater the distance from the centre ‘bulls-eye’ state, the greater the project variances (and the worse things are).

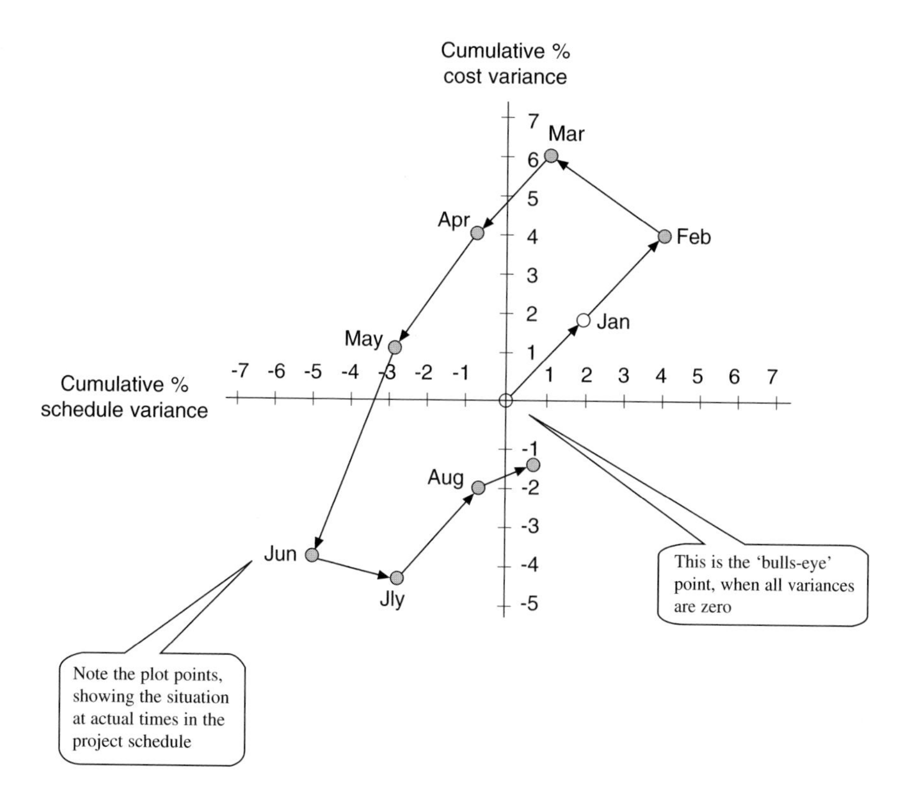

Fig. 9.6 A 'bulls-eye' project diagram

Project authority charts

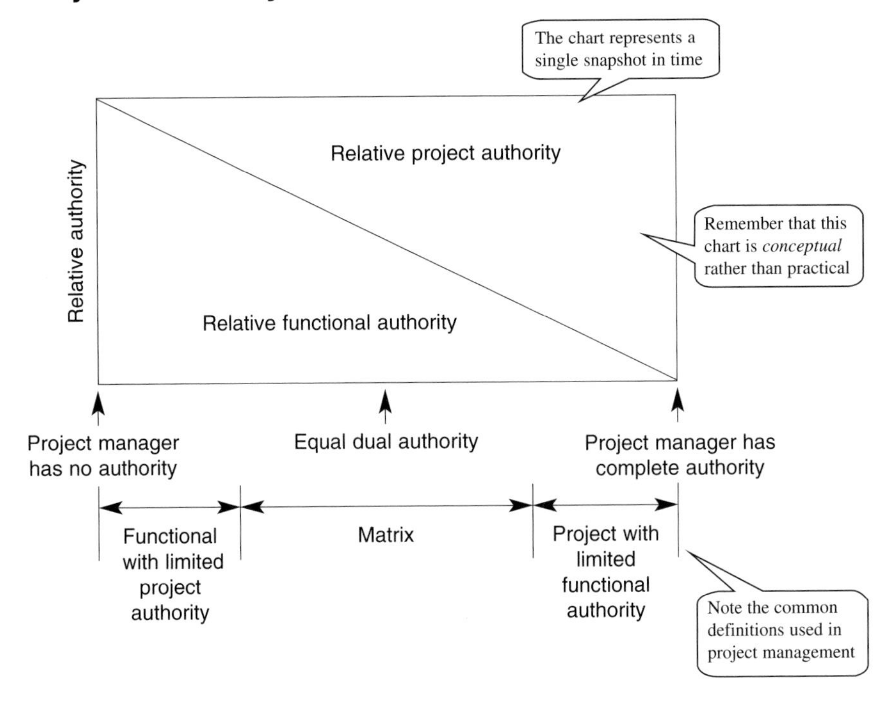

Fig. 9.7 A project authority chart

The purpose of these is to try to depict authority relationships within the organizational structure of the parties involved in managing a project (see Fig. 9.7). It is a conceptual rather than practical technique, often used in management-style presentations. Such charts are included in high-level management documentation circulated at the beginning of a project and again (later) if management problems occur.

PROJECT AUTHORITY CHARTS – KEY FEATURES

- The main objective is to solve the age-old problem of functional versus project management responsibilities as a project progresses.
- The chart represents only a snapshot in time in the project. Situations (and hence the chart) will change as the project progresses.
- They are conceptual tools. Do not expect them to be of any use in solving practical project management problems.

Project 'organograms'

'Organogram' is the name given to the type of organizational 'structure' diagram commonly used in project management. It is a purely hierarchical chart that tries to show who is in charge of the various parts and functions of the project. As with the project authority charts explained earlier, organograms are subject to change as a project progresses. This means that they are often only useful as 'guidance only' information rather than an accurate plan of how the project is being managed at any moment during its life-cycle. Figure 9.8 shows a typical organogram for a small project. Note the following features.

> **PROJECT 'ORGANOGRAMS' – KEY FEATURES**
>
> - A predominantly vertical, hierarchical layout with a few horizontal dotted lines indicating cross-disciplinary or inter-organization links.
> - The absence of any matrix-type structure. Organograms do not address the philosophy of 'matrix management'.
> - The necessary oversimplification of what are complex management relationships.

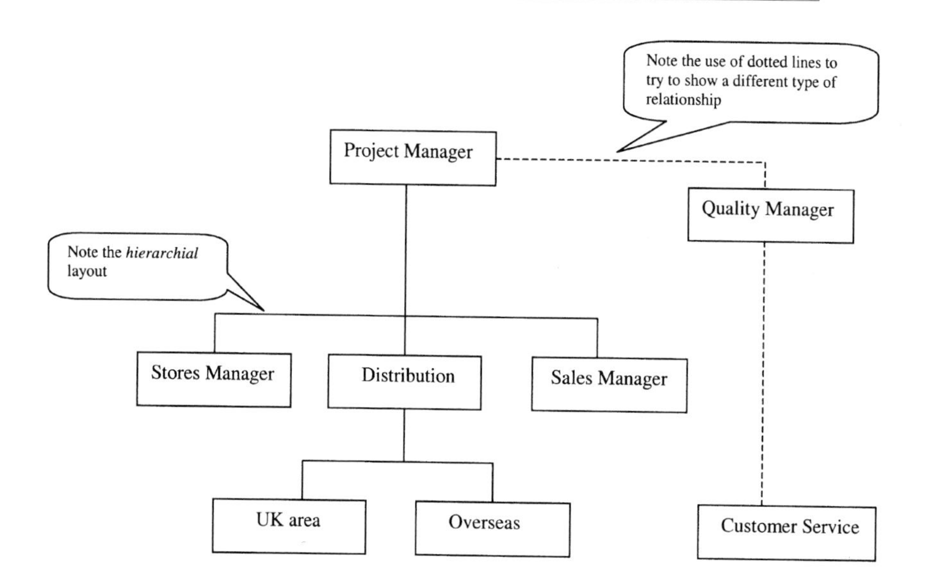

Fig. 9.8 A project 'organogram'

And then: Genesis – the 'viable system' organization model

What is it?

The viable system model (VSM) is an advanced graphical model of an organization. It is based on the principles of cybernetics which studies how organizations work, grow, and reproduce using analogies to biological organisms.

Its purpose?

To replace the view of the organization as a simple 'organogram' or hierarchical structure diagram which everyone knows has serious limitations in describing what actually happens in an organization.

The VSM Itself

Figure 9.9 shows the basic VSM. It consists of a series of different elements: autonomous 'producing' parts, internal control, external intelligence, and management policy parts all held together by links, checks, and balances. The model itself is three-dimensional, but represented on a two-dimensional page. This three-dimensional aspect means that the VSM can hold more information about an organizational system in a single diagram than almost any other type of model. It is a highly efficient way of representing organizational reality, which is why it is so complex. Note some more of the key features:

- the VSM is nested: each system contains, and is contained within, other systems of the same format;
- every part of the VSM is expandable: it can be increased (or decreased) in size and complexity to suit the organization being studied;
- the VSM is diagnostic: its purpose is to diagnose and describe problems in an organization, and so help find their solution.

The VSM shown in Fig. 9.9 represents one of the most advanced organization models that it is possible to present in the form of a single diagram. In terms of technical management information, it is difficult to find better.

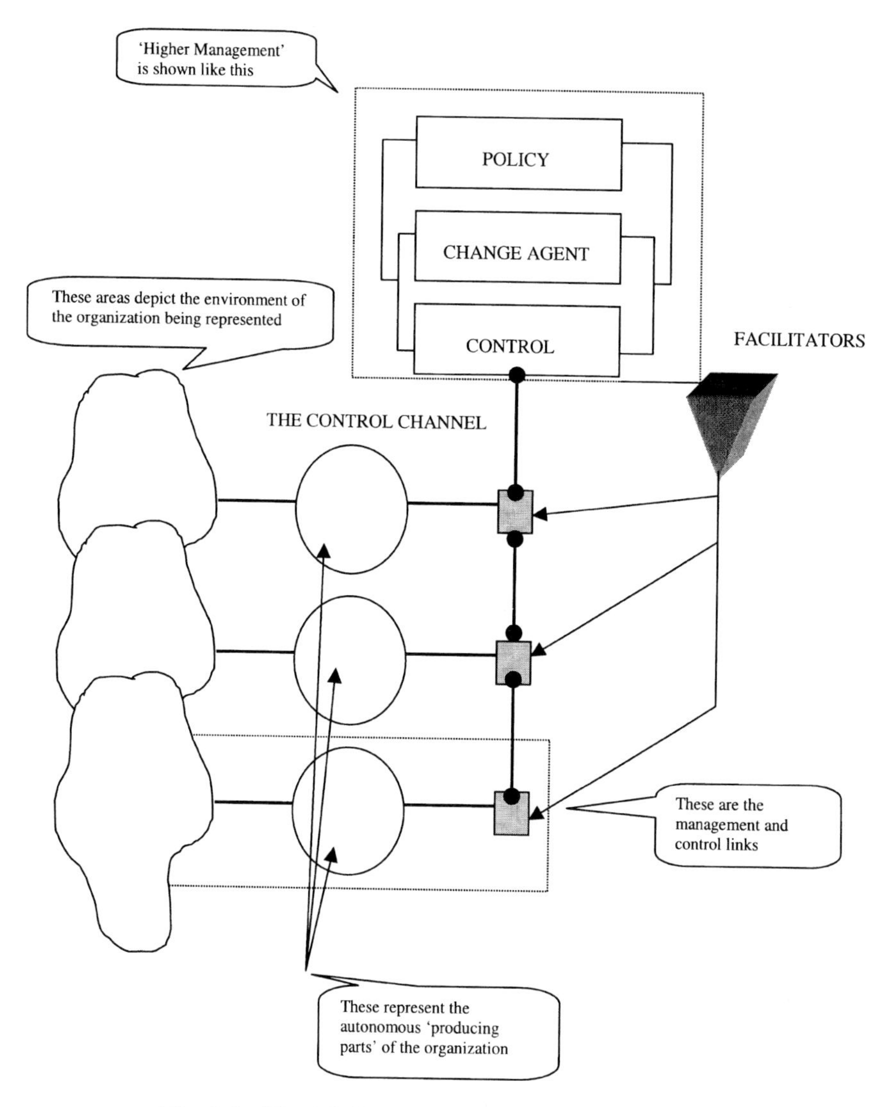

Fig. 9.9 The 'viable system' organizational model

Useful reference

Beer, S. (1979) *The Heart of Enterprise*, John Wiley and Sons Limited, Chichester, UK.

Chapter 10

Statistical Information

Introduction – presenting statistical information

Statistical techniques provide a reliable means of describing accurately all manner of information about technical subjects. Most technical disciplines have a quantitative background and the presentation of data relevant to these disciplines is often best handled by statistics. At its most basic, statistics is simply about dealing with groups of numbers, choosing various ways to present information and the inferences that are contained within. The techniques of presenting statistical information fall into several well-defined groups. The use of these is common to just about all technical disciplines.

STATISTICAL DATA PRESENTATIONS DIVIDE BROADLY INTO SIX CATEGORIES SHOWING:

- averages and means;
- probabilities;
- central tendency;
- variability;
- distribution shape;
- relationships between groups of things.

We will look in this chapter at some of the common ways to present statistical information about technical subjects. They will be related where possible, to the formal statistical techniques that lie behind them.

Boxplots

These are sometimes referred to as 'box and whisker' plots, and are a way of showing the spread of a set of data readings. Figure 10.1 shows how the basic 'box' is constructed from a batch of data readings.

BOXPLOTS – KEY FEATURES

- The central 'box' encompasses two 'quantities'.
- The lines from each side of the box (the whiskers) show the location of the other data points.
- The relative sizes of the box and its whiskers give a useful impression of how the batch of data is distributed.

There are several possible variations of the standard 'quartile' boxplot. Lines with arrowheads are conventionally used to represent deciles, which omit the extreme upper and lower 10 percent of the data values (a common technique for large batches of data of say, 100 000+ readings where the extreme values are not that important).

Using boxplots for comparisons

The standard quartile boxplots are also a useful way of presenting comparisons between batches of data.

'COMPARISON' BOXPLOTS – KEY FEATURES

- Note how the inter-quartile ranges (the boxes) in Fig. 10.1 are similar lengths. This indicates data that warrants comparison, i.e. not widely differing sets of data for which a comparison would be meaningless (like comparing sizes of apples and bananas, for example).
- The shaded portions allow a quick visual assimilation of the quartile ranges.
- The relative lengths of the whiskers allows a quick visual comparison of the distribution of the datasets, without looking up the individual readings.

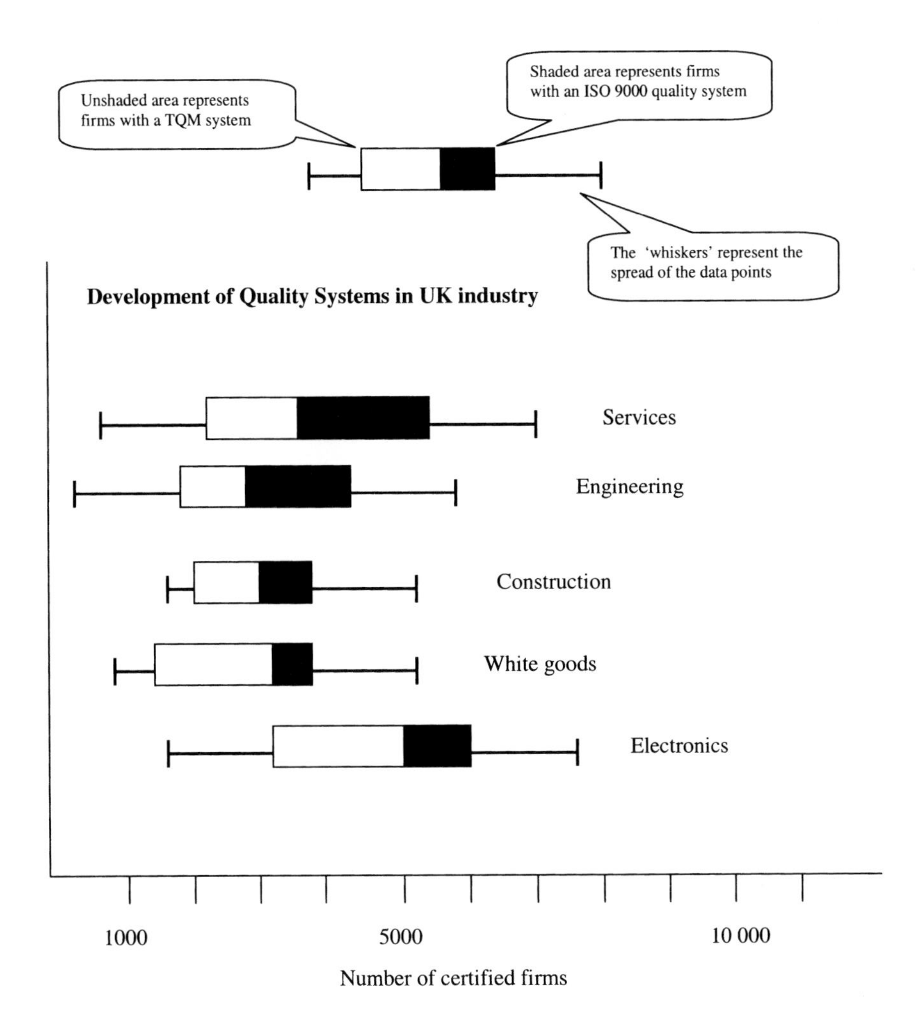

Fig. 10.1 A boxplot graph

Traditional histograms

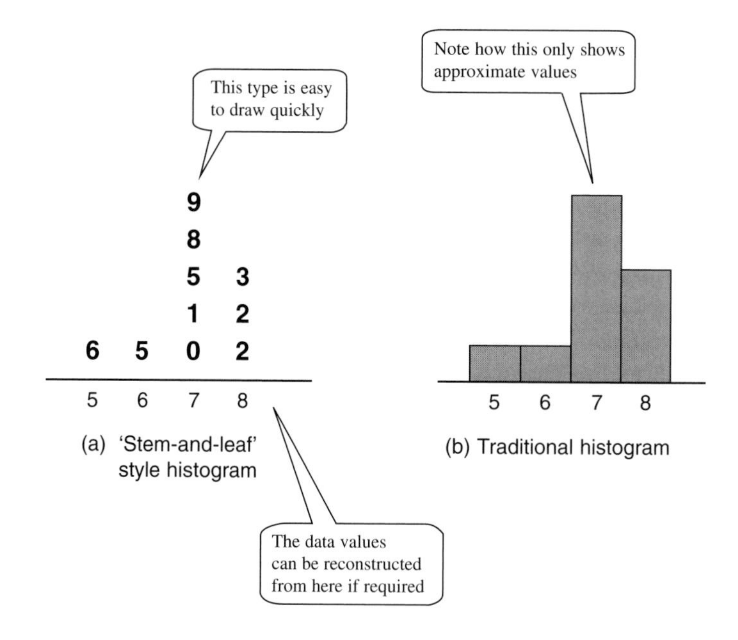

Fig. 10.2 The two types of basic histogram

The histogram is the most basic visual method of presenting statistical data. It shows which data values in a set are common or rare, roughly where the middle values are, and how well spread out the data are. It holds no information, however, on the order of any of the data readings. Figure 10.2 shows the two basic types of histogram.

Figure 10.2(a) is a 'stem and leaf' type histogram. Its advantage is that the original data set can be reconstructed back from the histogram.

Figure 10.2(b) is the traditional bar chart histogram (see also Chapter 2) It conveys the same information as (a), but the data values cannot be 'read back'.

Cumulative frequency graphs

A cumulative frequency graph, as the name suggests, shows cumulative information about a set of data readings. This is particularly useful when it is necessary to know the percentages of data readings less than or greater than a certain value. Figure 10.3 shows a typical example. Note how quartile categories can be easily shown. It is convention to show cumulative frequency on the vertical axis.

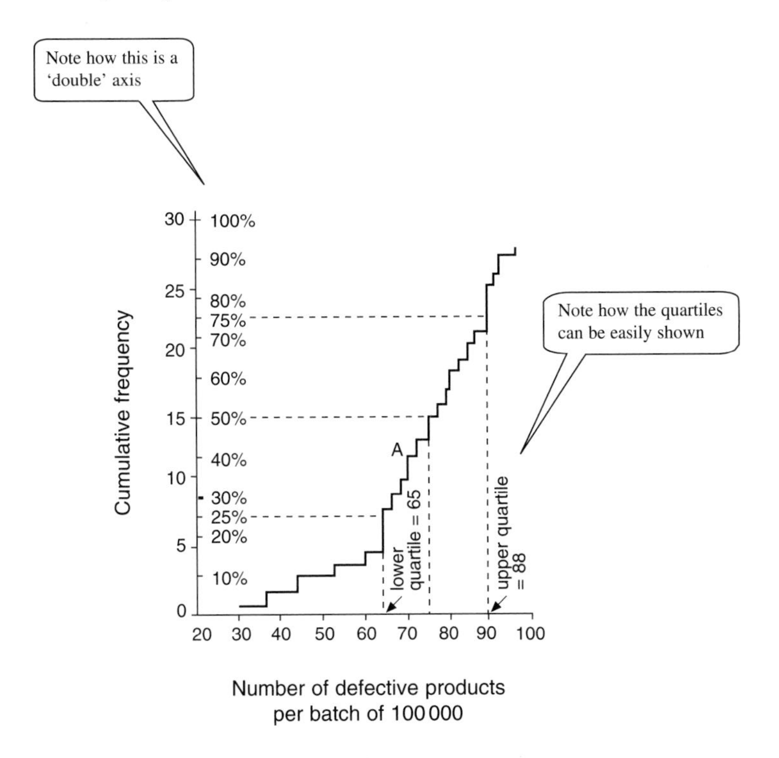

Fig. 10.3 A cumulative frequency graph

Frequency histograms

A frequency histogram is simply a pictorial way of presenting information that could be shown in tabular form.

FREQUENCY HISTOGRAMS – KEY POINTS

- The bases of the histogram columns correspond to the 'internal ranges' of the data set.
- The areas of the columns represent the frequency with which the data readings occur.
- Remember that histograms describe the 'shapes' of data that have been measured on a continuous number scale. This is different from bar charts which depict separate categories of data.
- The width of the histogram columns do not have to be the same (although they normally are).

Figures 10.4(a) and (b) show examples of the basic frequency histogram and a version in which the midpoints of each column are joined by a curve. This gives a characteristic which can then be displayed alone (without the columns).

Useful reference

http://franz.stat.wisc.edu/~rossini/courses/intro-biomed/text/Frequency_Graphs.html

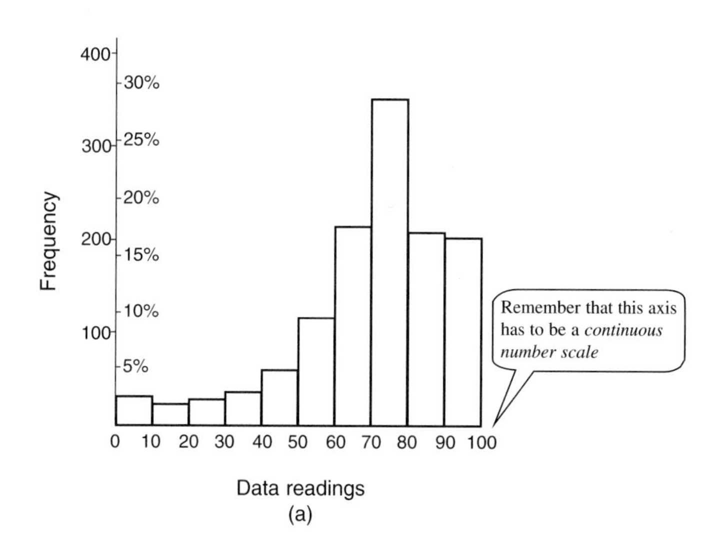

(a)

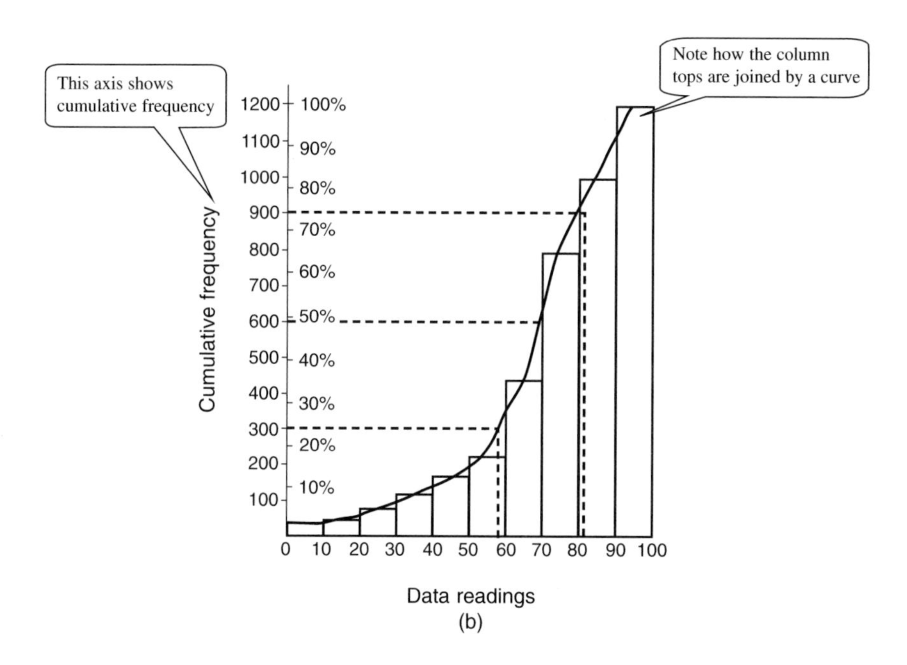

(b)

Fig. 10.4 Two types of frequency histogram

Probability models

The concept of probability has relevance to many technical disciplines. There are numerous instances in technical analysis, design, and manufacturing where events cannot be absolutely defined and it is necessary to resort to statistical techniques to predict when a component is likely to fail, or the likelihood that a medicine will work, or similar.

Probability is a topic which allows us to understand, calculate, and compare various risks inherent in the technical world. It is also an aid to drawing sensible conclusions from technical data. The key concept for presentation purposes is that of probability distributions; these are effectively models of reality which help in the process of drawing conclusions. Technical disciplines use several well-accepted distributions, each with their own presentation features. The most common one for presentation purposes is the normal distribution.

The normal distribution

Also known as the Gaussian distribution or the 'Z' distribution, this is shown as a bell-shaped curve with a hump in the middle and a tail on either side. It has a specific form of curvature which follows a mathematical equation. Figure 10.5 shows a general example.

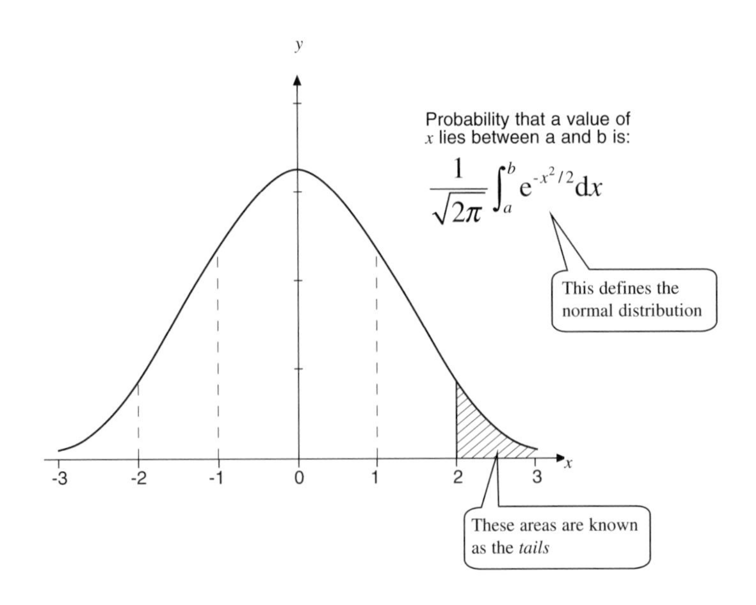

Fig. 10.5 The normal distribution

THE NORMAL DISTRIBUTION – KEY FEATURES

- Purpose. It is a model of reality.
- Content. The distribution assumes a large data set.
- Definition. A normal distribution can be defined from knowing only two things: the mean and the standard deviation (σ).
- Calculating probabilities. The objective of the distribution (any distribution) is to enable you to calculate probabilities. Figure 10.5 shows how this is done.

Bivariate plots

Bivariate data – what are they?

Bivariate data are data that consist of two related paired groups of numbers, each of which may have different units of measurement. They are commonly presented in graphical form and are subject to advanced statistical techniques that look at the relationship between two sets of properties.

The simple bivariate plot

You may also see this referred to as a 'scatterplot', 'x–y plot', or 'two-dimensional plot'. The points are plotted against simple x, y axes (Fig. 10.6 shows an example.

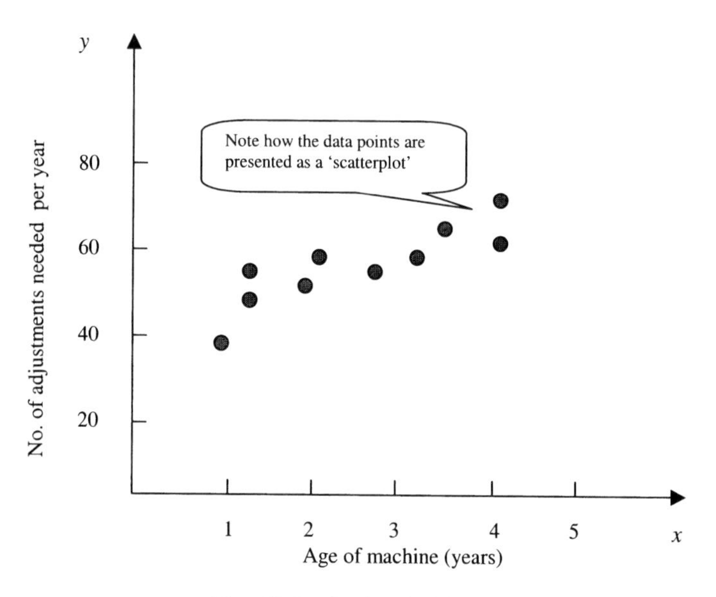

Fig. 10.6 A simple bivariate plot

> **SIMPLE BIVARIATE PLOT – KEY FEATURES**
>
> - The plot shows how the age of a machine and the number of adjustments required are related to each other.
> - It also shows something of the distributions of age and number of adjustments separately.
> - The shape of the distribution of points is starting to take on a structure. In this example, for instance, we can see that the number of adjustments generally increases with age, i.e. a pattern is starting to emerge.

Other bivariate plots

In many technical disciplines, bivariate data plots will indicate some relationship between the two properties x and y of the data readings. Figure 10.7(a) shows how the plot looks when the data exhibit a linear trend, i.e. when y increases in a straight line relationship with x. Figures 10.7(b) and (c) show two separate types of non-linear trend in which x and y do not follow a linear relationship. Note also the existence of more random data readings, termed 'outliers'. Data presentations like this are common in real-world technical situations.

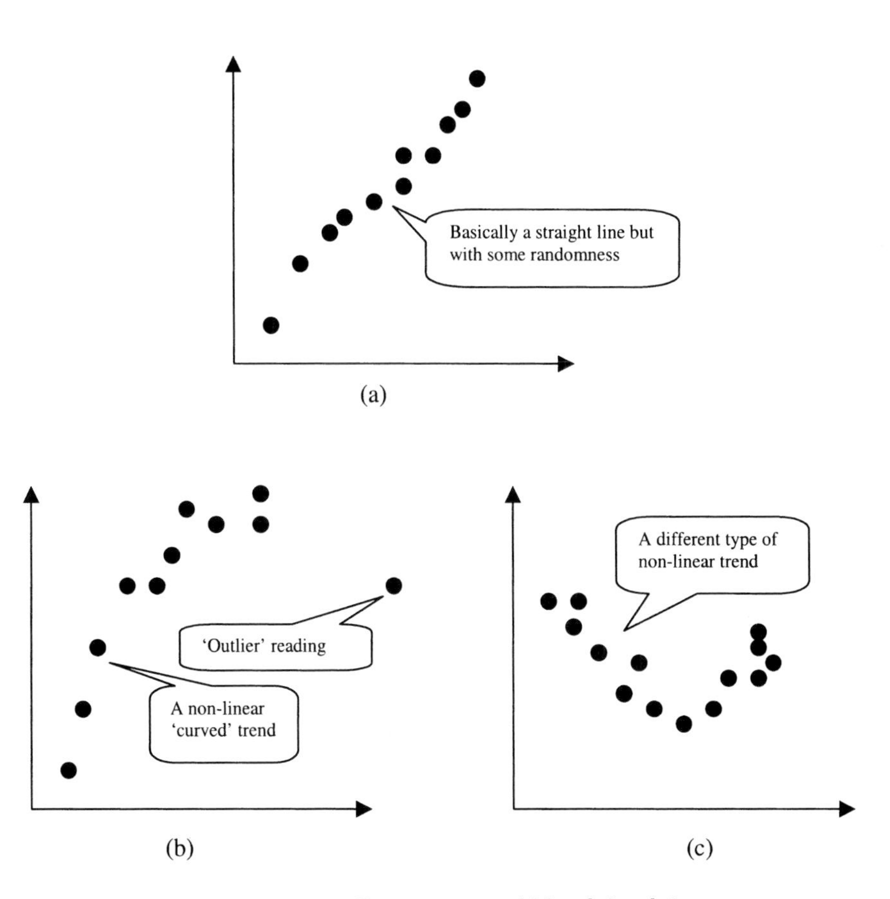

Fig. 10.7 Different types of bivariate plot

Correlation graphs

Correlation is a measure of how strongly the two factors x and y (of a set of bivariate data) are associated with each other. There is a firm mathematical basis in the form of a correlation coefficient (r), a number between -1 and +1 which acts as the measure of linear association.

Figures 10.8(a)–(d) show typical presentations for various values of r.

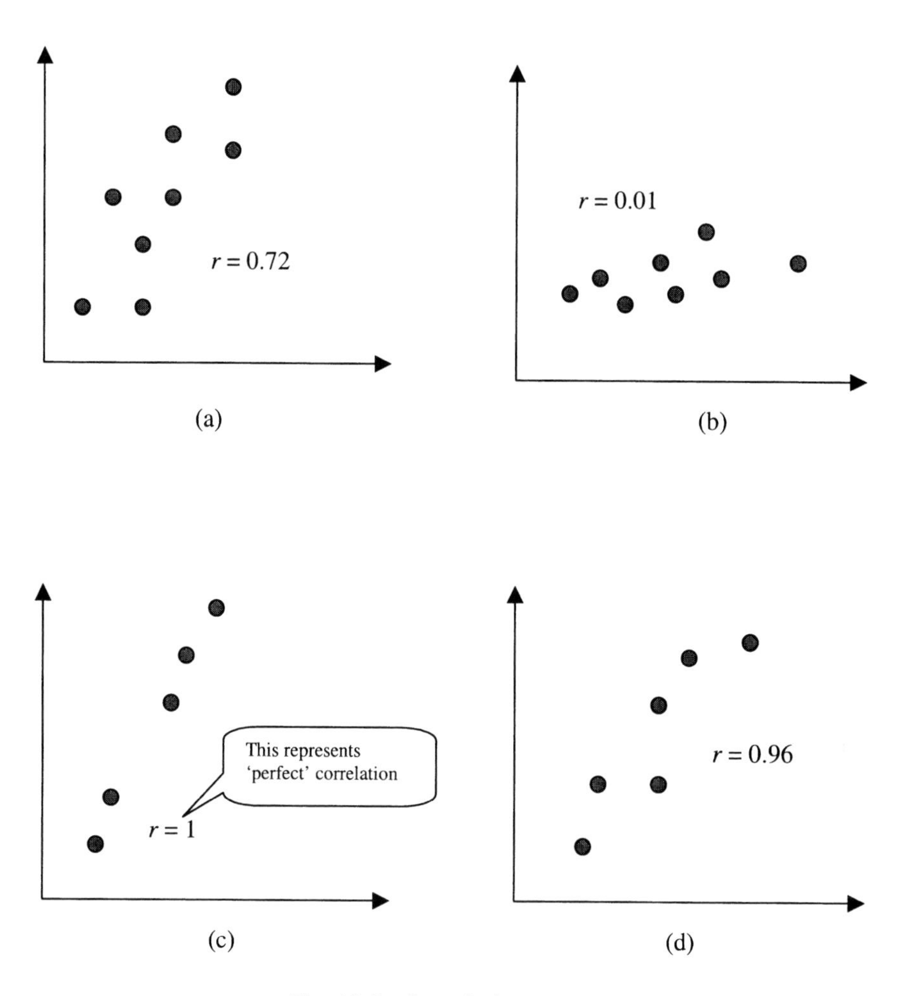

Fig. 10.8 Correlation graphs

Regression analysis graphs

Regression analysis is simply a way of indicating the basic relationship between two variables x and y (in a bivariate plot) by drawing a line passing through the data points. This line is called the 'best-fit' or 'regression' line and is drawn either by eye or using a mathematical technique termed 'least squares analysis' (easily handled by a computer programme). As with correlation, the regression technique uses mathematical coefficients to describe the 'line'. Figure 10.9 shows the way this is presented.

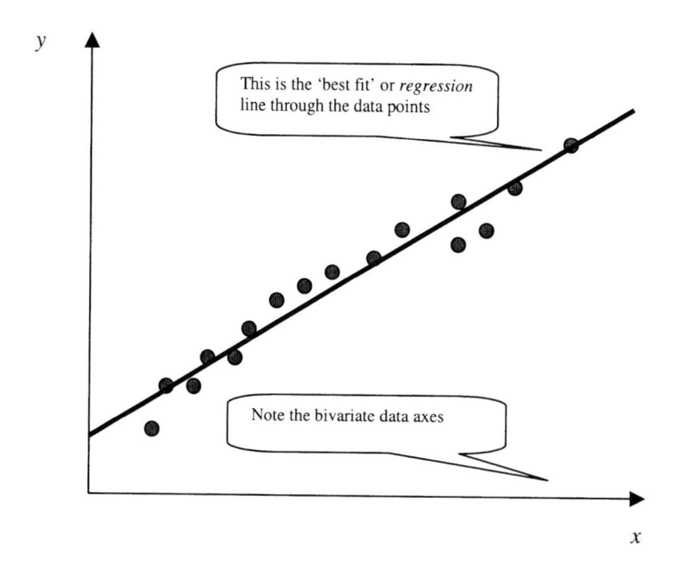

Fig. 10.9 A regression analysis graph

Statistical process control (SPC) charts

SPC is a technique widely used in the manufacture of mass produced mechanical, electrical, and consumer products. It is a systematic tool for analysing the state of the manufacturing process (i.e. to see whether it is in 'equilibrium' and so capable of consistently producing items to the necessary size, surface finish, quality, etc. Two specific types of control charts are used: range charts and average charts.

Range charts

Range charts show the upper and lower control limits (of dimension or whatever) and the desirable value or 'central line'. The actual data from individual manufactured items are plotted on the charts, giving an impression of whether the process is 'under control'. Figure 10.10(a) shows examples, along with the standard forms of terminology used.

Average charts

These show the control limits for average readings measured from batches of the manufactured products. They are structured in a similar way to the range charts, but use different symbols. Figure 10.10(b) shows typical examples.

SPC CHARTS – POINTS TO REMEMBER

- The objective of control charts is to show whether an industrial process is 'under control' or not.
- There are two types: range charts and average charts.
- Expect a lot of specialized terminology and abbreviations; they are all referenced in published technical standards.

Useful reference

http://www.west.asu.edu/tqteam/stat.html

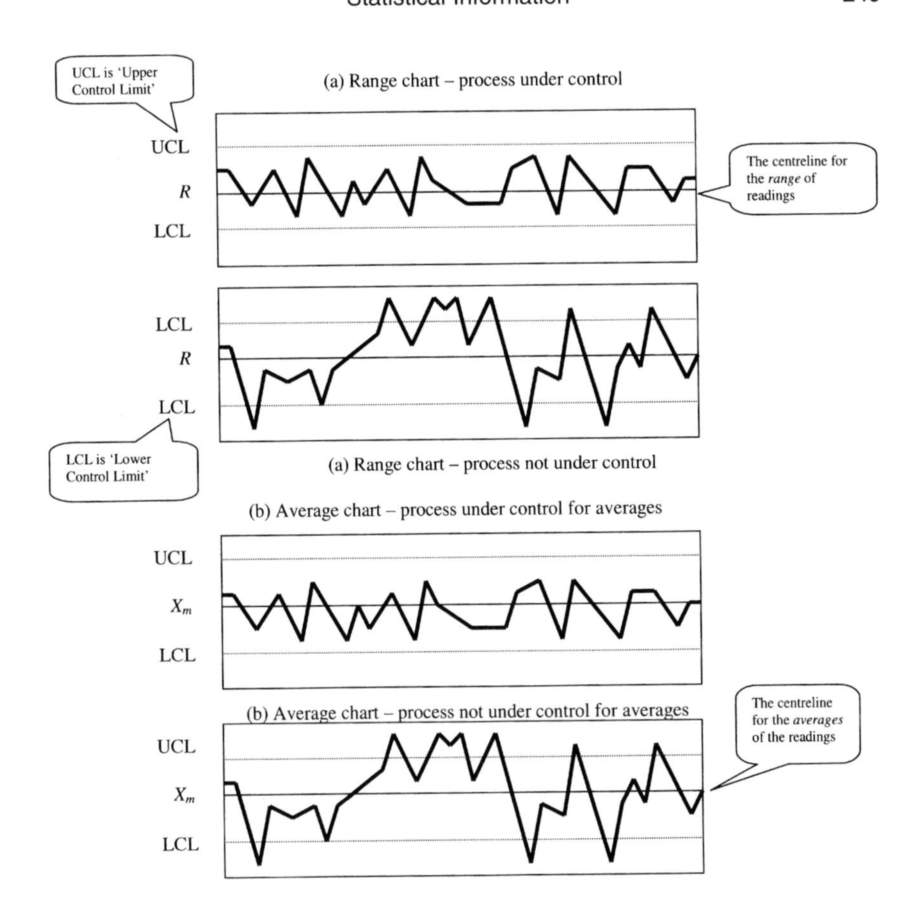

Fig. 10.10 Statistical process control (SPC) charts

Surface texture charts

Surface texture charts are a practical way of presenting technical information, but also have a robust statistical basis. They are used to convey detailed and accurate information about the quality of a machined or ground surface (normally of a metal). Most rely on the analysis of a distribution around the centre or 'average' line. Figure 10.11 shows two typical charts.

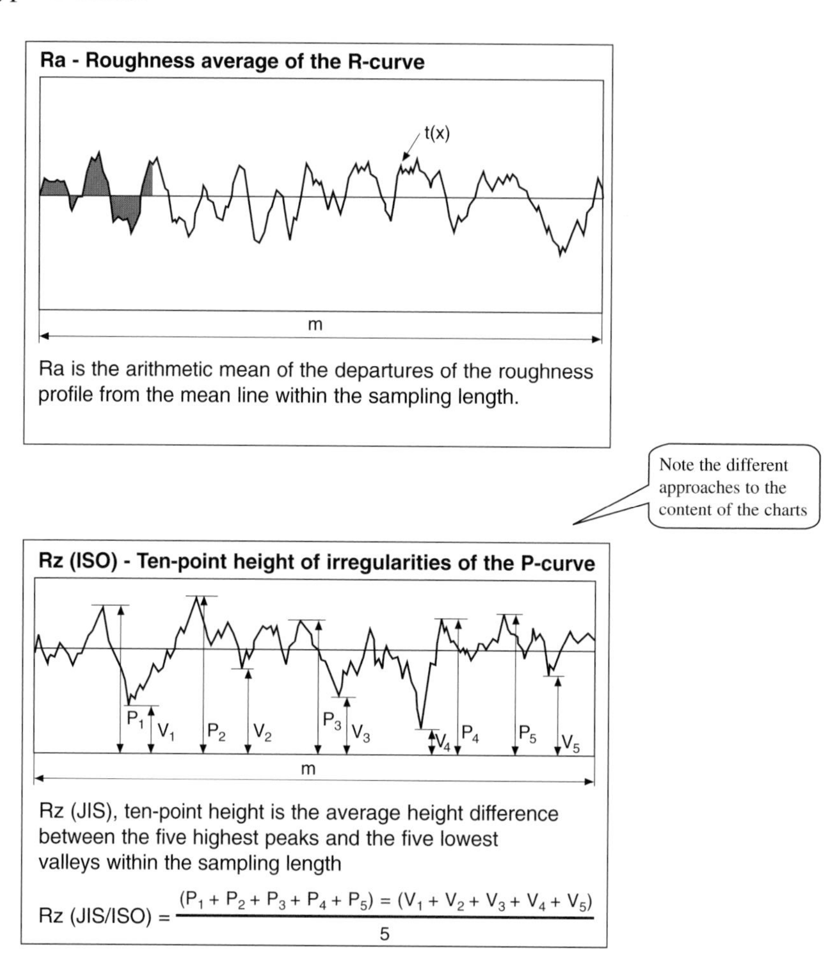

Fig. 10.11 Surface texture charts

Chapter 11

Thermodynamics/Fluid Mechanics

Introduction – representing 'conceptual' quantities using diagrams

Thermodynamics and fluid mechanics are two separate, but related, technical disciplines that make extensive use of technical presentation methods. While it is untrue to categorize these disciplines as 'conceptual' (they are in fact heavily quantitative), it is fair to say that many of the ways that technical information about thermodynamics and fluid mechanics is presented have a conceptual content to them. This is because they rely on convention, i.e. accepted ways of presenting complex ideas in a way that people can understand.

Both thermodynamics and fluid mechanics are based around mathematical laws, axioms, and rules. So, for accurate representation of technical information, mathematical expression is the best method. However, mathematics alone, while accurate, cannot always be assimilated by non-technical readers, so a variety of visual ways to present the information is also needed. These visual presentation methods are often relevant to both thermodynamics and fluid mechanics, owing to the analogies between the two subjects.

This chapter looks at a selection of presentation methods, with particular emphasis on those used in the practical application of thermodynamics and fluids technology, rather than pure, more theoretical activities.

Heat transfer modes

Different types of heat transfer processes are known as modes. The three main ones are conduction, convection, and radiation. In thermodynamics it is often necessary to represent these heat transfer modes within the context of complex diagrams showing heat transfer in and around bodies and engineering systems. The three modes are conventionally shown in the schematic format shown in Figs 11.1(a), (b), and (c). Note the use of standard terminology with its subscripts and superscripts.

SHOWING HEAT TRANSFER MODES – KEY POINTS

- Conduction is shown by the existence of a temperature gradient 'line' across the section, and two temperatures T_1, T_2 (Fig. 11.1(a)).
- Convection: note how the moving fluid is represented as a band of arrows (Fig. 11.1(b)). The surface temperature is T_s and the convective heat flux (q'') is specified.
- Radiation is denoted by a surface flux value (q) and a surface emissivity (ε). Note how this is a schematic representation of a complex heat transfer condition (Fig. 11.1(c)).

Useful reference

http://www.does.org/masterli/e39_e40.html

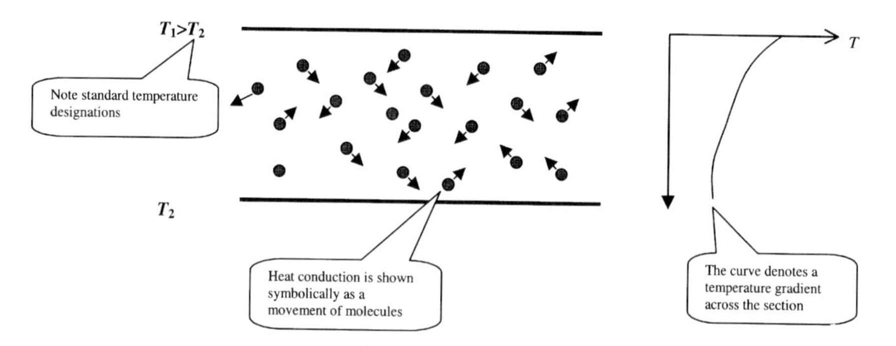

Fig. 11.1(a) How to show conduction heat transfer

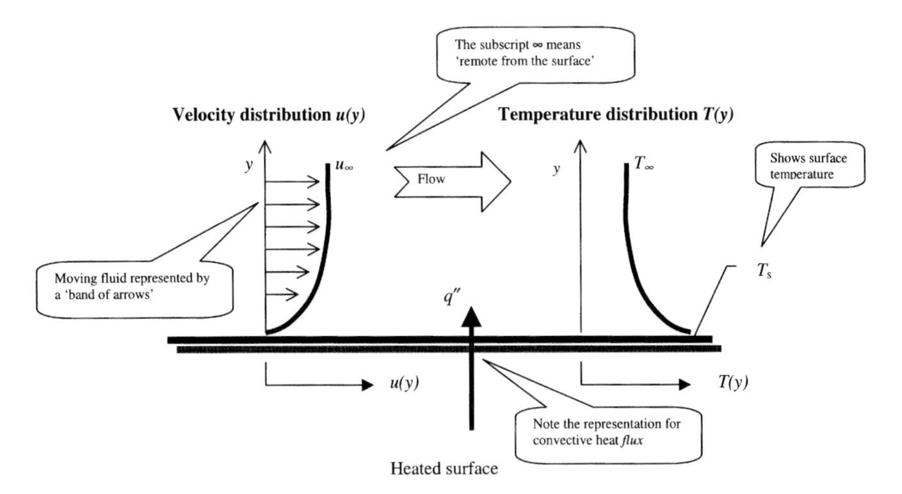

Fig. 11.1(b) How to show convection heat transfer

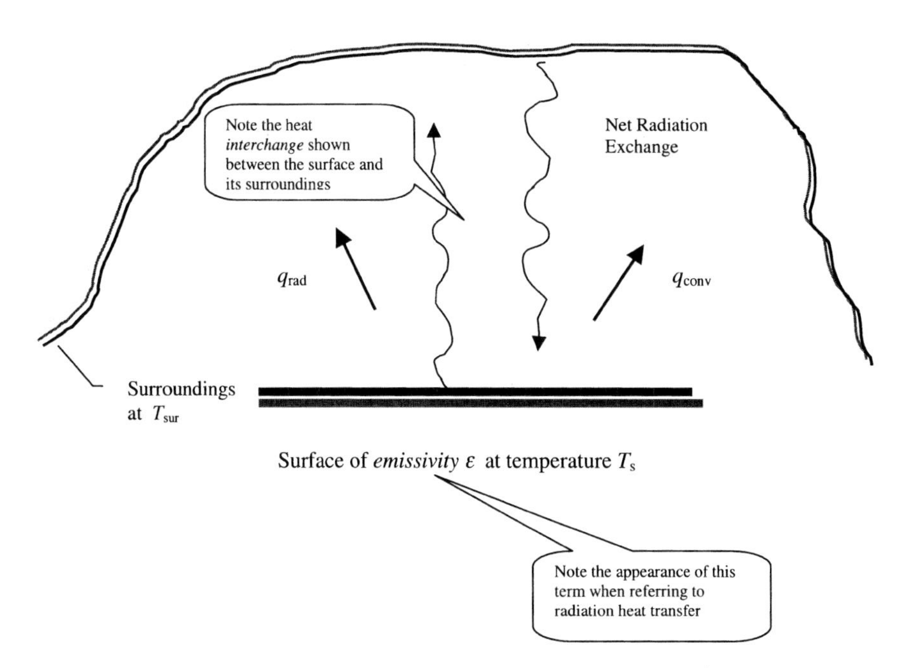

Fig. 11.1(c) How to show radiation heat transfer

One-dimensional conduction diagrams

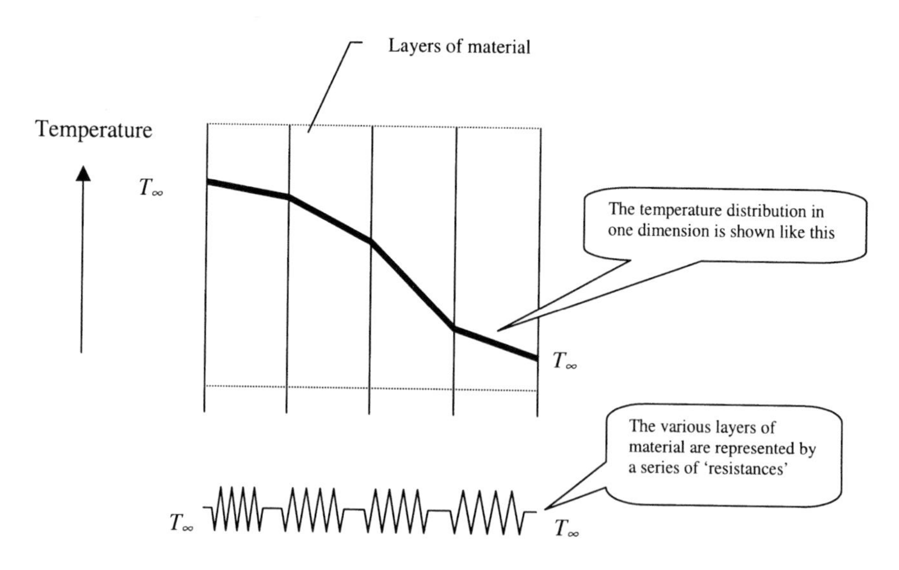

Fig. 11.2 How to show 1-D conduction

More complex conditions of one-dimensional conduction are presented as shown in Fig. 11.2. This shows how the temperature distribution across the wall of a composite surface is depicted. Note the following features:

- the line graph showing temperature distribution;
- the electrical resistance diagram analogy.

Two-dimensional conduction diagrams

Problems in thermodynamics are rarely one-dimensional; in many cases a minimum of two-dimensional analysis is needed. A presentation method used to give a good first estimate of temperature distribution in a 2-D object is the 'flux plot'. It is a hand-drawn or computer-generated network of perpendicular isotherms and heat flow lines. The result is a network of curvilinear squares which is then analysed to infer the temperature distribution in the medium, and the heat transfer rate. Figure 11.3 shows a simple example.

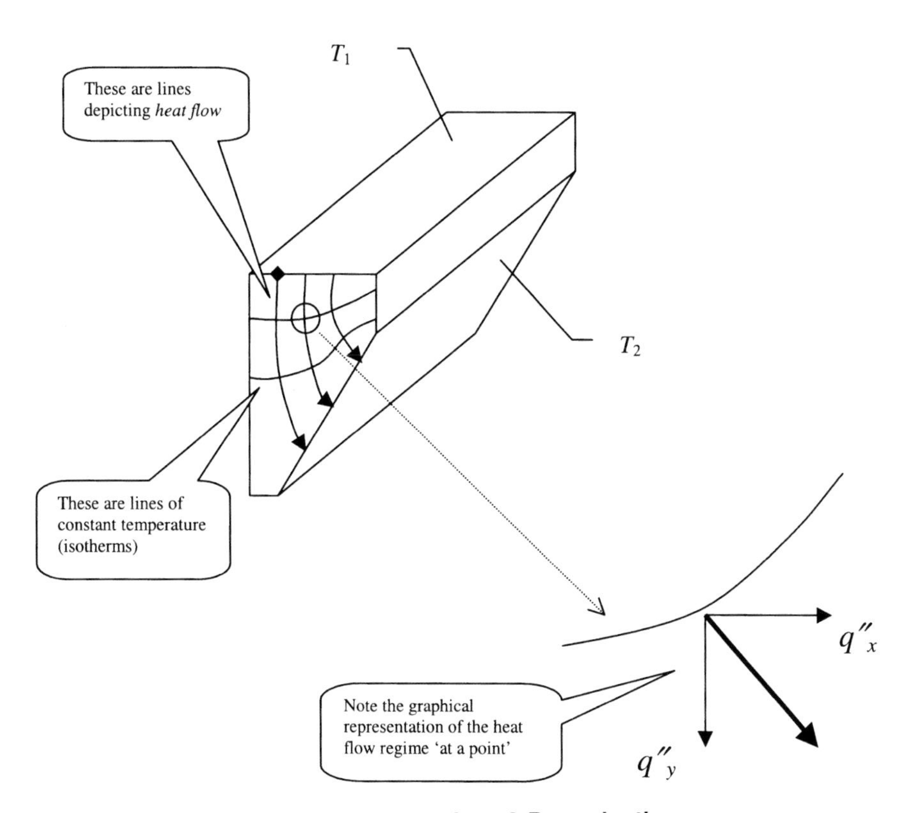

Fig. 11.3 How to show 2-D conduction

Boundary layer diagrams

Technical and design problems in thermodynamics and fluid mechanics often involve the consideration of boundary layers. There are various types of boundary layer: thermal boundary layers are formed by temperature differences, concentration boundary layers are the result of varying concentrations of the process fluid, and velocity boundary layers are caused by varying velocities. All of these types can be shown graphically, as well as algebraically. Figures 11.4(a), (b), and (c) show the conventional way in which the three types are presented. Note the key features.

BOUNDARY LAYER DIAGRAMS – KEY FEATURES

- All the diagrams show a boundary layer profile expressed as a band of arrows of varying lengths.
- Note the free stream region. This is the region that is considered to be unaffected by the existence of the surface.
- Boundary layer diagrams are nearly always simplified by representing the surface as a flat plate. This is an approximation, to stop the analysis getting too complicated.

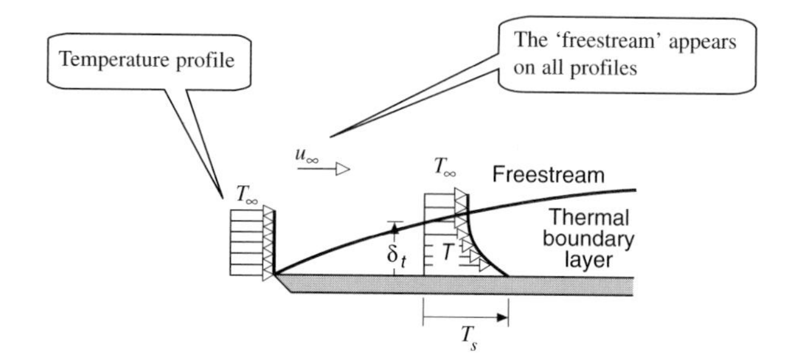

(a) A 'temperature' boundary layer

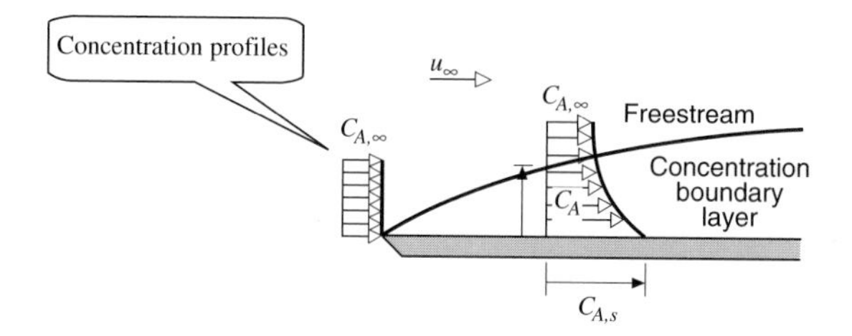

(b) A 'concentration' boundary layer

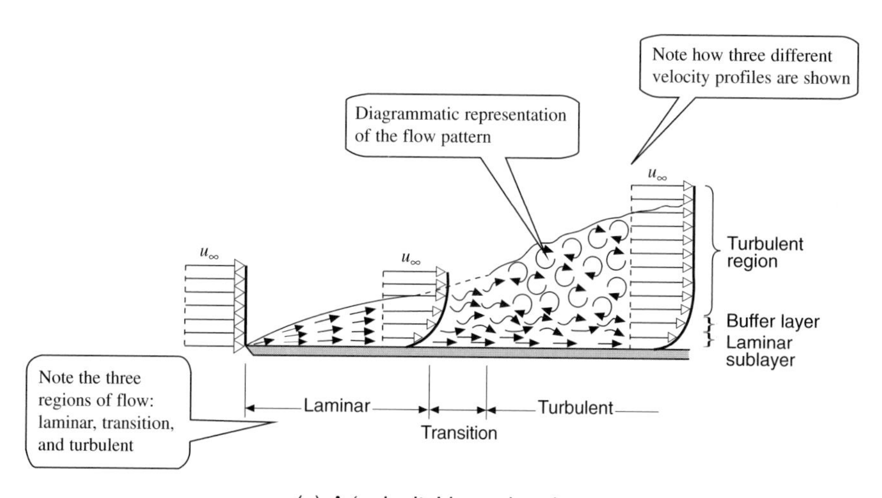

(c) A 'velocity' boundary layer

Fig. 11.4 Diagramatic views of boundary layers

Internal and external flow diagrams

Heat transfer design problems tend to separate into those involving 'internal' flow through a body and those to do with 'external' flow around a body. They have different thermodynamics and hydrodynamic considerations, and hence their graphical presentations look different. As for boundary layers, internal and external flow diagrams are simplified by representing the geometry as a pipe or cylinder – this is done even in quite complex designs. Figures 11.5 and 11.6 show typical presentations. Note the key features.

> **INTERNAL FLOW DIAGRAMS – KEY FEATURES**
>
> - Note the simplified 'pipe' flow area (Fig. 11.5).
> - It is normal to show four separate regions of flow: entrance, inviscid, boundary, and fully developed.
> - Look at the emphasis placed on the shape of the flow profile. It is shown at a minimum of three locations, to show how the flow develops.

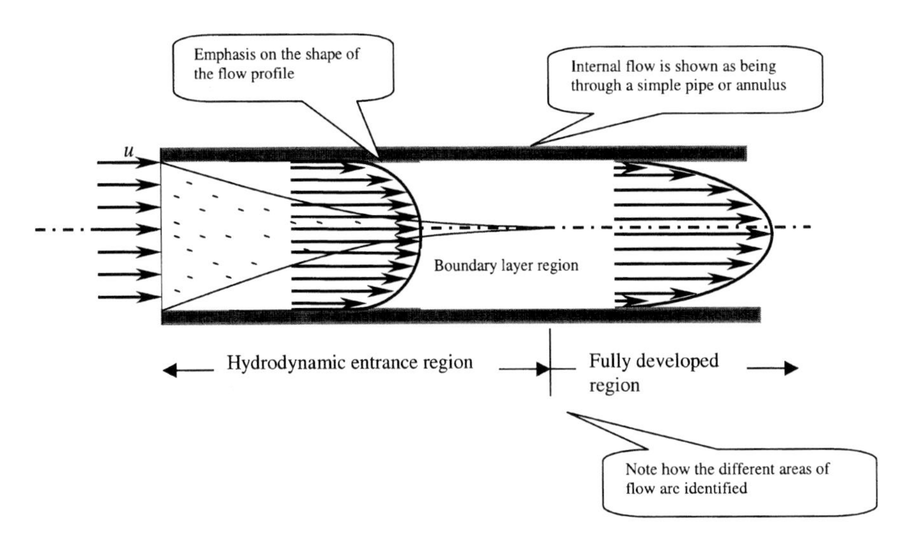

Fig. 11.5 A typical 'internal flow' diagram

EXTERNAL FLOW DIAGRAMS – KEY FEATURES

- Note the simplifications; plate and cylinder shapes act as the basic 'references' for more complex surface geometries (Fig. 11.6).
- Look at the conventions for showing wakes and vortices.
- Note the additional mathematical expressions needed to describe the external flow region.

Boundary layer and separation on a cylinder

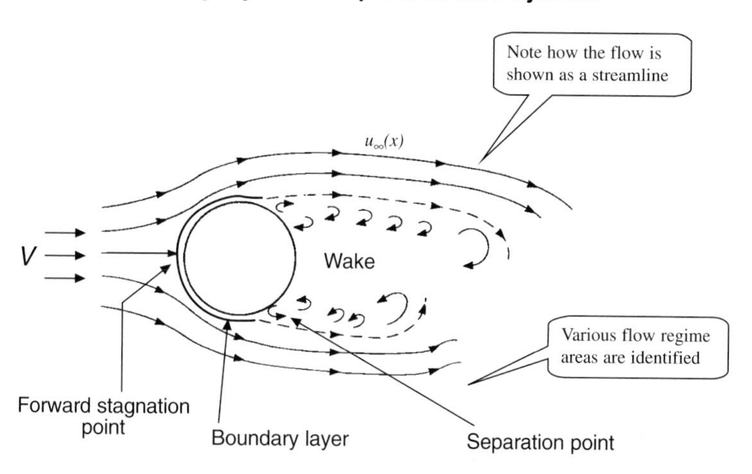

Velocity profiles over a surface

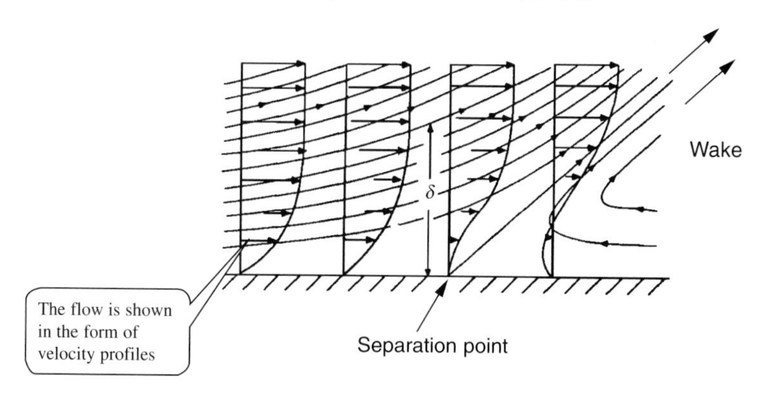

Fig. 11.6 Two ways of showing 'external flow'

Thermal distribution – schematics

Schematic diagrams showing heat flow through regions and surfaces are used in many practical design applications. They are easily understood by semi-technical readers. Note the key features.

> **THERMAL SCHEMATICS – KEY FEATURES**
>
> - Thermal 'heat' flow is shown visually with thick lines representing greater heat flow.
> - Percentages are shown also; these are necessary to give some quantitative accuracy to the diagram.
> - The presentation is purely schematic; there is no attempt to represent thickness and distances to scale.

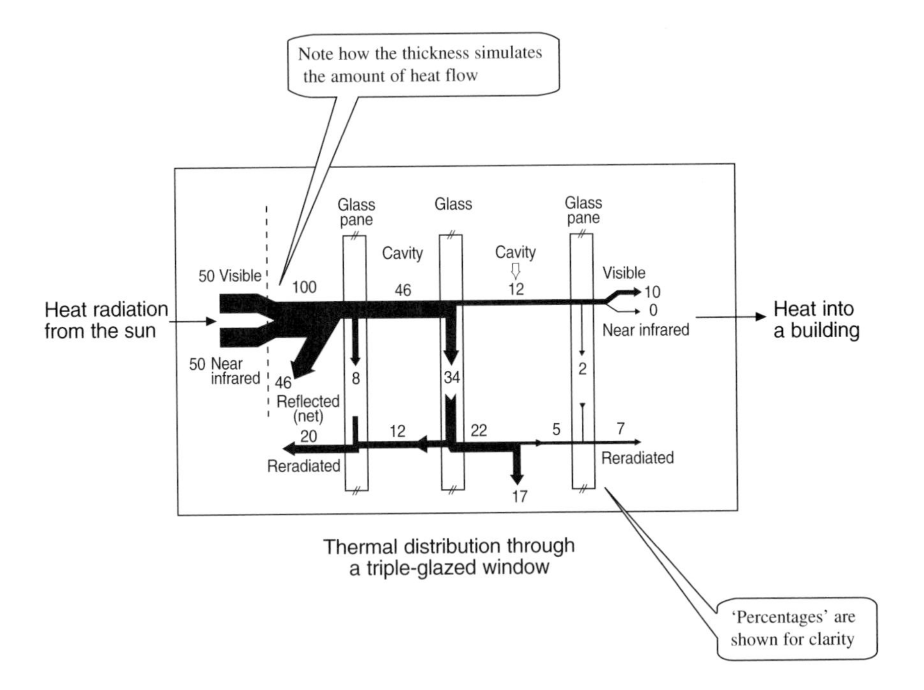

Fig. 11.7 A 'thermal distribution' schematic

Thermal performance – schematics

A thermal performance diagram is a more complex example of the thermal distribution schematic. It is also schematic and provides a global view of the thermal performance of a complete system. Typical applications include:

- showing heating/cooling loads in buildings;
- optimizing the overall thermal profile of process plants;
- showing flow distributions in process plants, i.e. using the same principle to show fluid distributions (a different objective but the same type of schematic presentation).

Figure 11.8 shows a typical example for the thermal performance of a building. Note how all this information could have been presented in tabular form, but would not be so easy to assimilate.

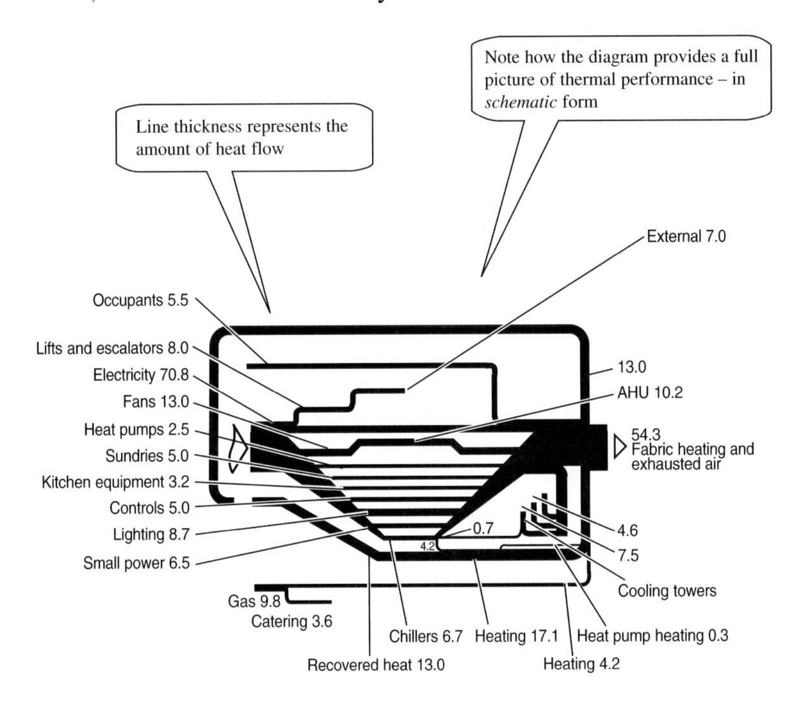

Fig. 11.8 A 'thermal performance' schematic

Fluid–pressure diagrams

These are mainly used during the design process for fluid equipment or fluid-based machine elements such as bearings, gearboxes, and fluid couplings. The diagrams consist of a geometric view of the component on which is superimposed a pressure distribution curve (Fig. 11.9).

> **FLUID–PRESSURE DIAGRAMS – KEY FEATURES**
>
> - They show the distribution of fluid pressure (not necessarily the individual values, which may be indeterminable).
> - The diagrams are partly geometric and partly schematic. The components are drawn to scale but the pressure distribution curves are generally a representation only, drawn to an arbitrary scale.
> - Note the use of clearly labelled dimensions (particularly for angles). This is important as it allows the expression of precise mathematic relationships later in the design process.

Oil pressure distribution in a journal bearing

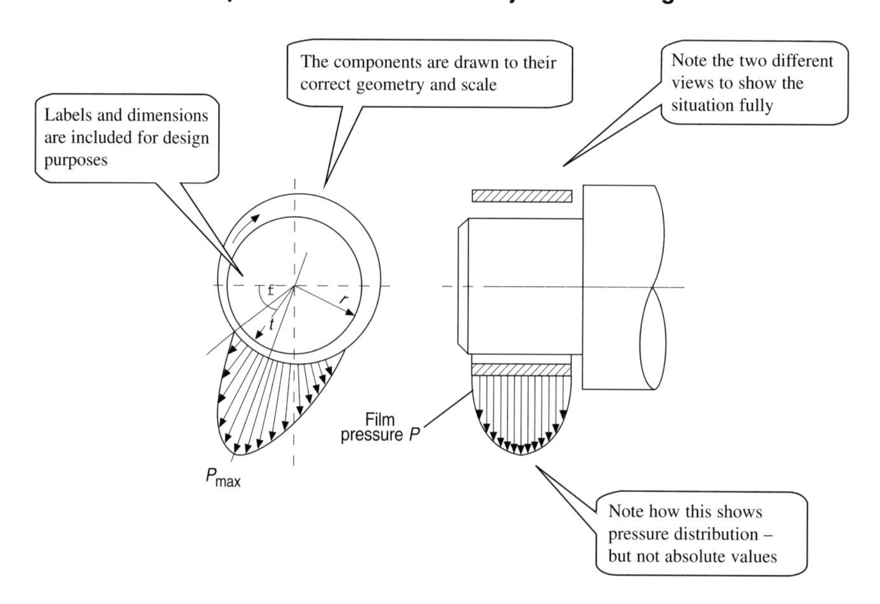

Fig. 11.9 A fluid–pressure diagram

Pressure–volume and temperature–entropy

The essence of applied thermodynamics is the study of the relationship between heat, work, and the properties of systems. Inherent in this is the need to show changes in the working fluid contained in a system. Diagrams showing this are used continually in applied thermodynamics, the most common ones being pressure–volume (p–v) and temperature-entropy (T–s) diagrams.

P–v diagrams

P–v diagrams are shown in a simple x–y axis configuration, as illustrated by the example in Fig. 11.10. Note the key features:

P–v DIAGRAMS – KEY FEATURES

- The points are connected by curves. Separate unconnected points have no real significance on a p–v diagram.
- Solid line curves denote an (ideally) reversible process. Dotted lines are used if the process is irreversible.
- Separate points are shown on the curve. These represent states of the working fluid and are labelled with numbers or letters.

The effect of heating a liquid

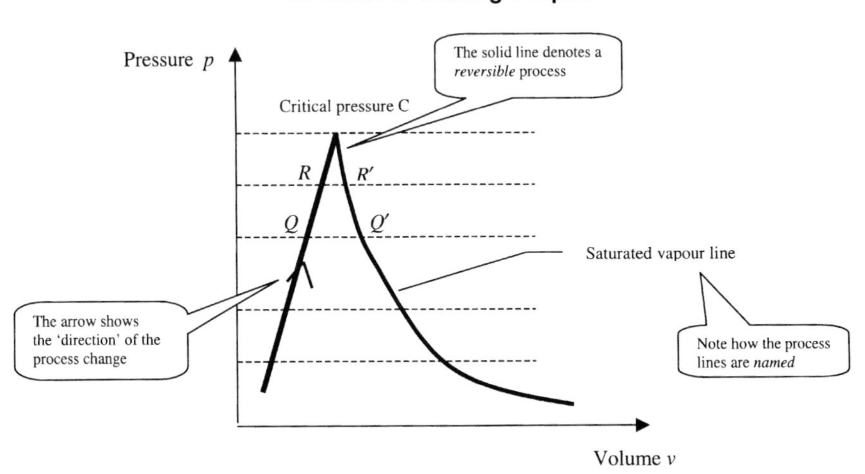

Fig. 11.10 A typical p–v diagram

T–s diagrams

These are used to show entropy, a property which follows from the second law of thermodynamics. Entropy is also a property represented by a reversible adiabatic process and is conventionally shown on a simple two-axis line graph (see Fig. 11.11). Note the key features.

> **T–s DIAGRAMS – KEY FEATURES**
>
> - The main objectives of *T–s* diagrams is to show *changes* in entropy rather than individual 'point' values.
> - The area under a *T–s* curve is often shaded to represent the amount of heat flow.
> - Additional (solid) curves showing lines of constant pressure are often superimposed on to the *T–s* diagram. Some also show constant volume lines (dotted curves).

A *T–s* diagram for vapour

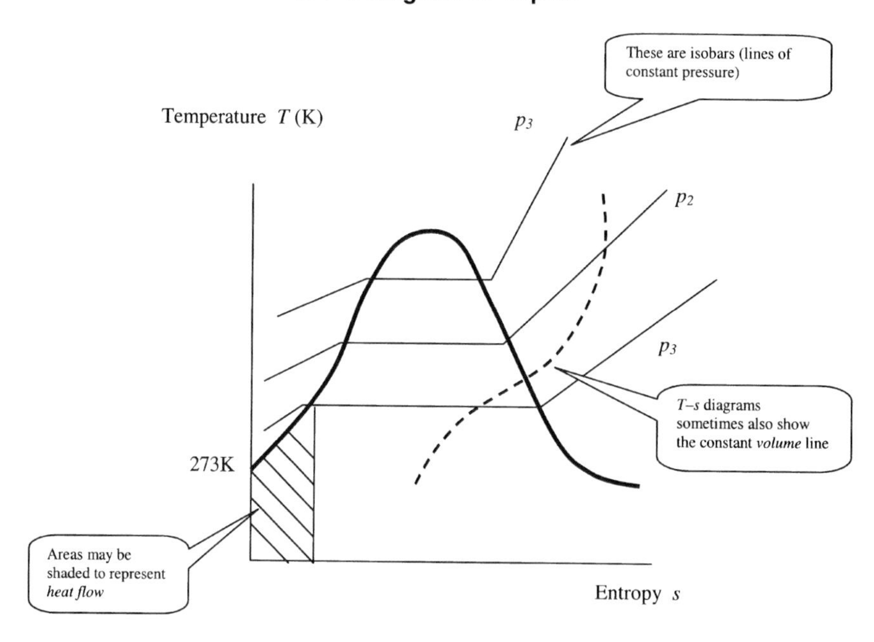

Fig. 11.11 A typical *T–s* diagram

Equi-flow diagrams

There are two types of these, used to show graphically important practical situations in fluid mechanics. They are both complex presentations, often prepared by computer.

Equi-vorticity/streamline diagrams

These have the specific purpose of showing flow 'streamlines' and areas of equal flow vorticity in the flow regime of an incompressible fluid. They are a visualization of complex equations and numerical methods. Figure 11.12 shows an example for the flow through a pipe expansion. Note the key features.

EQUI-VORTICITY DIAGRAMS – KEY FEATURES

- For symmetrical arrangements the diagram can be split to show equi-vorticity lines in one half and flow streamlines in the other.
- The relative dimensions of the flow area must be shown – the flow lines are specific to the geometric arrangement.
- Parameter values of the lines are usually shown.

Non-compressible flow through a sudden expansion

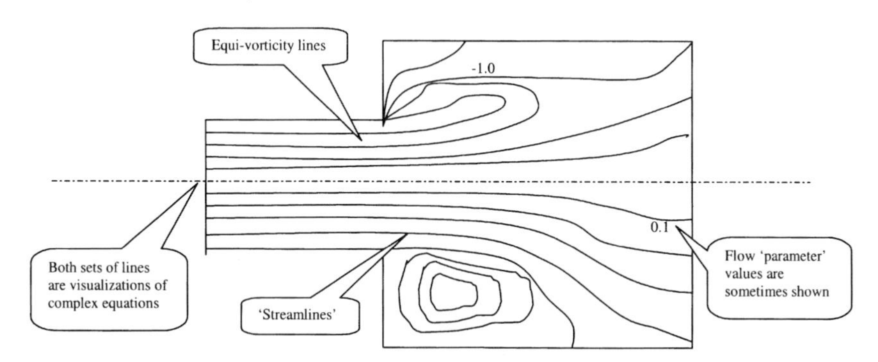

Fig. 11.12 A typical 'equi-flow' diagram

Equi-mach diagrams

This is a method of showing velocity distribution in a compressible fluid such as air. Its most common use is the visualization of shockwaves in supersonic aircraft and projectiles (Fig. 11.13). The data relating to the diagram are produced by complex differential equation methods, making it suitable only for computer analysis.

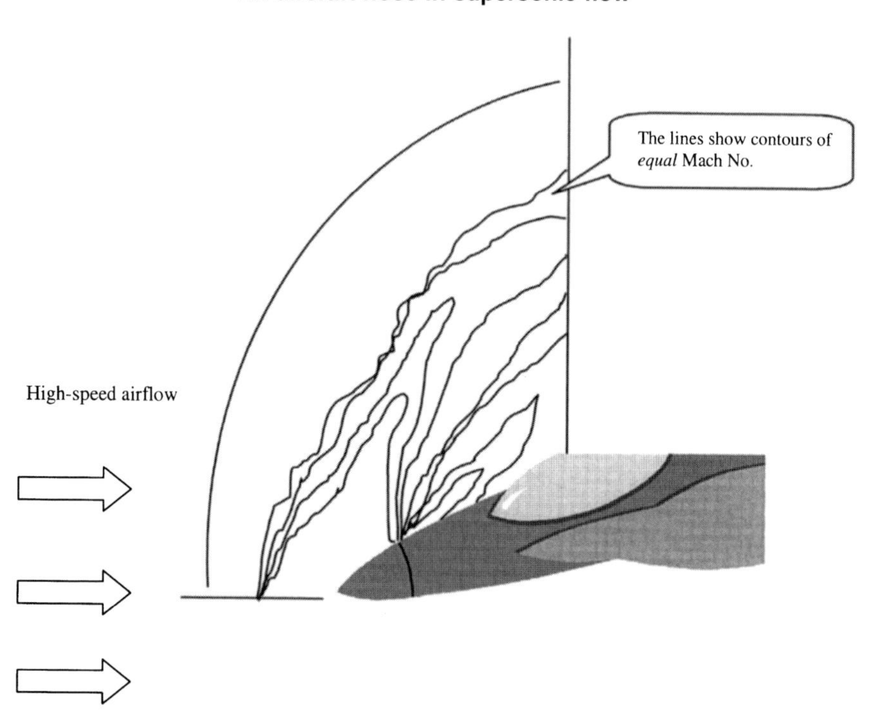

Fig. 11.13 A typical 'equi-mach' diagram

Flow visualizations

Flow of air and some other fluids cannot be seen by the naked eye, so various methods are used to present information in a visual form. Figure 11.14 shows several of the more common types.

Visualized image analysis

This is a digital technique in which streamlines are recorded by a camera, converted to binary codes and processed into velocity 'bits' as shown (Fig. 11.14(a)).

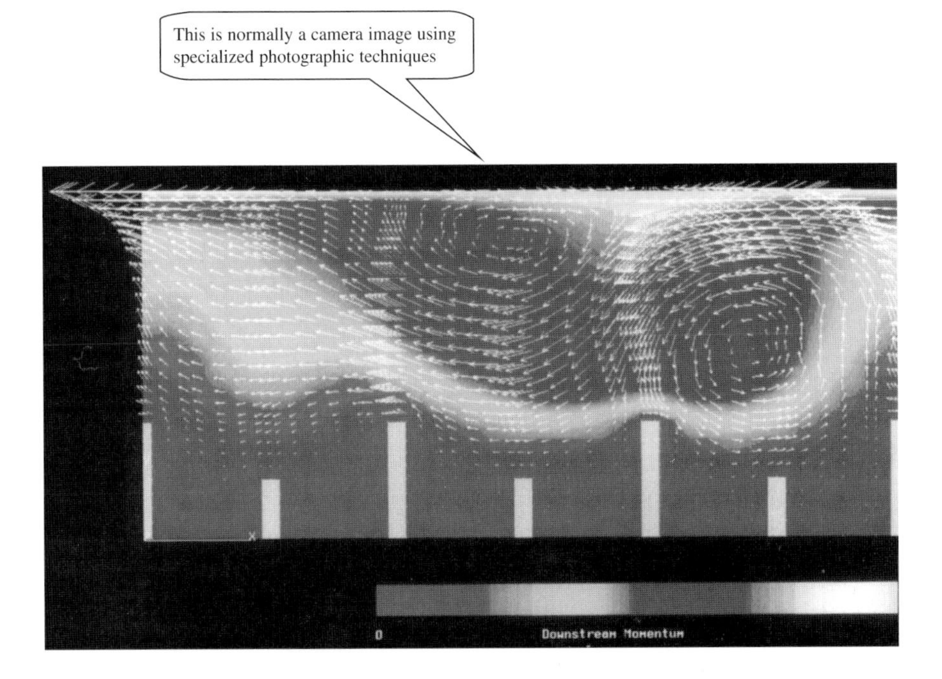

Fig. 11.14(a) Showing flow using visualized image analysis

Mean velocity vector displays

In this technique, flow data are analysed by computer and the output presented as in Fig. 11.14(b). The size and direction of flow velocity are represented as vectorial arrows.

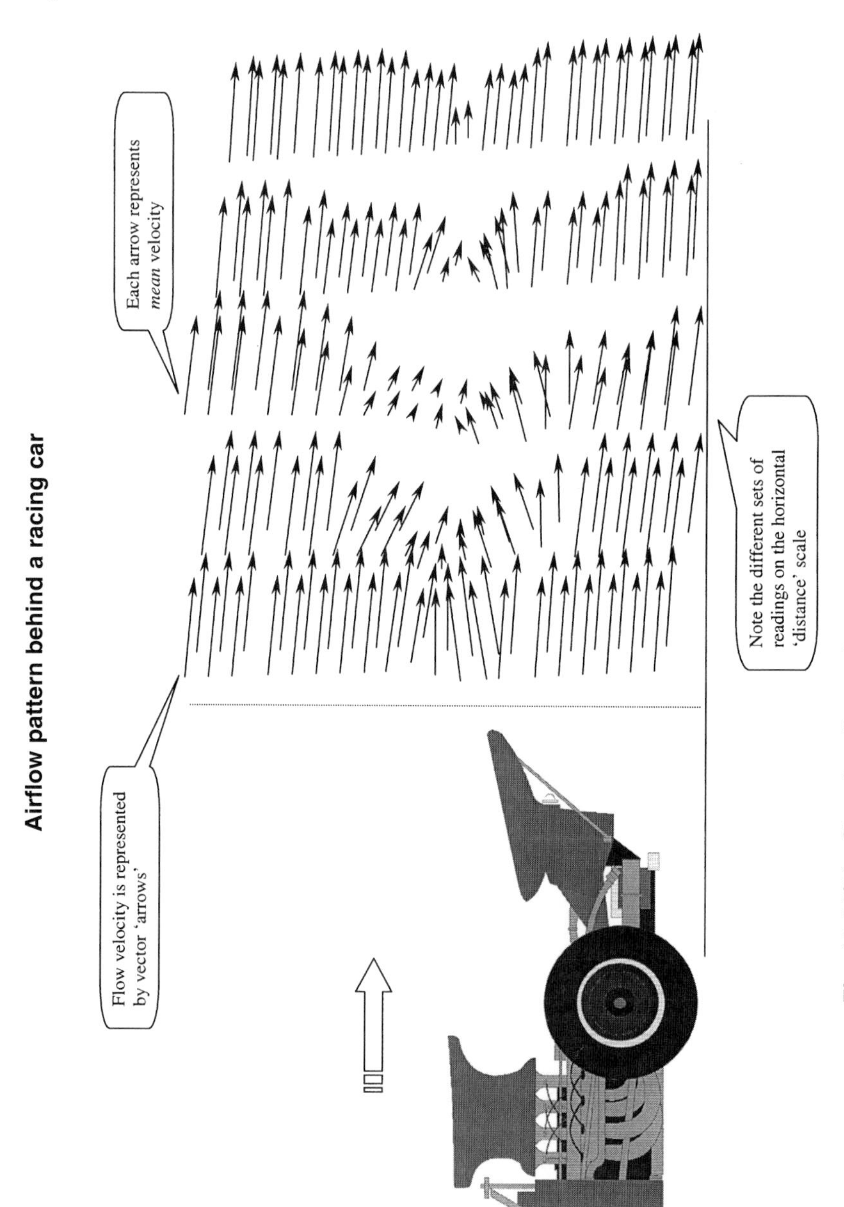

Fig. 11.14(b) Showing flow using a mean velocity vector display

Radiating flow displays

This is a relative of the vector display method and is often used in acoustics. The diagram comprises two parts: a side 'section' view showing vector information and a complementary view showing how the flow radiates (see Fig. 11.14(c)). This is a highly visual method that can be adapted to use colour displays and advanced 3-D computer graphics.

The acoustic power flow from a guitar

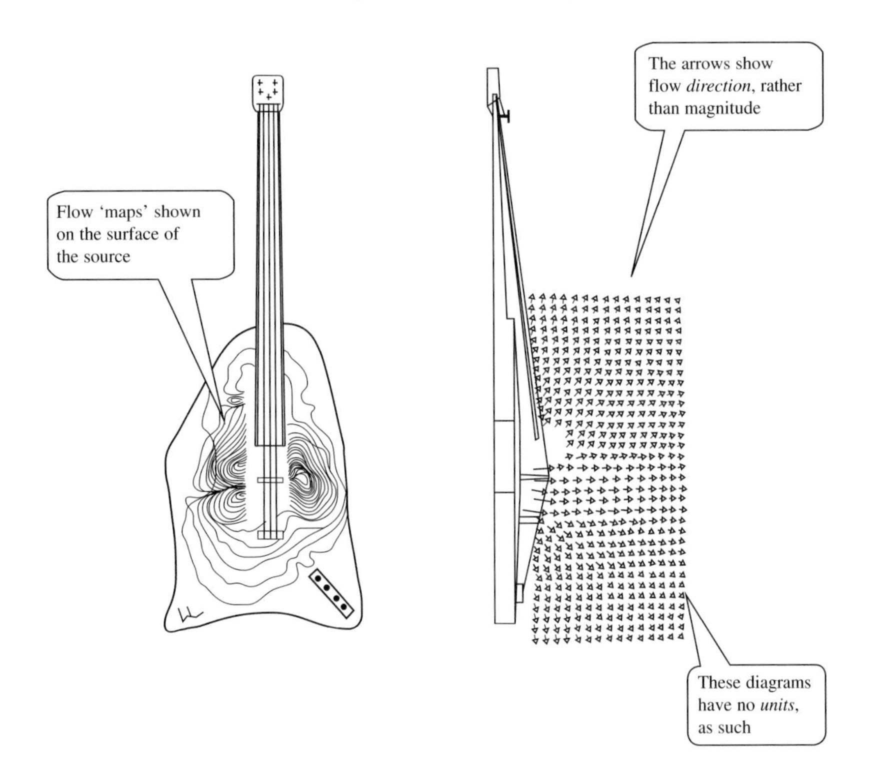

Fig. 11.14(c) A typical 'radiating flow display' diagram

Useful references

http://www.wake.tec.nc.us/SVG/svg.p2.html
http://noodle.med.yale.edu/

Chapter 12

Process Engineering

Introduction – process drawings and diagrams

The disciplines of process engineering can be broadly defined as those branches of engineering that deal with flow systems. Within this designation lies a wide variety of different types of system containing chemicals, steam, hydraulics, and pneumatics. All are linked together by monitoring, instrumentation, and control systems.

Despite the variety of the subject, most process engineering disciplines involve similar sorts of drawings and diagrams. Much the same as we saw in Chapter 3, these are the result of experience and convention. Their format is almost universal, with only minor differences between industries and countries. As with engineering drawings, the main objective of process drawings is that they provide a unique and absolutely unambiguous description. This means that several different 'levels' of drawing are needed, in order to accommodate all the necessary technical detail. This chapter looks at the various types, and shows how the technical content is divided up between them. Content rather than style is the most important part of process diagrams. Styles do vary, depending on the type of drafting technique or software used.

PROCESS ENGINEERING DIAGRAMS – SOME GENERAL POINTS

- They follow similar formats and conventions across industry boundaries.
- Expect industry-specific symbols and methods of labelling.
- Remember that process engineering diagrams (like engineering drawings) come in sets, each providing a part of an overall picture.

Process schematics

Process schematics, often known as 'process flow sheets' show the overall arrangement or 'scheme' of a process. They show the existence of elements such as process vessels, rotating equipment, fluid and solid transfer equipment, and the pipework and ductwork that connect them all together. Because they are schematic, these diagrams do not attempt to show the physical size or layout of the system components, only how they are connected together. In most industries, the process schematics are the means by which engineers 'read' the process, and they are one of the first stages when a new process is being designed from scratch.

In large industrial applications such as chemical, pharmaceutical, or power plant, a project will have several sets of process schematics divided up into 'plant area' groups. These are conventionally numbered as areas 100, 200, 300, etc. Figure 12.1 shows a typical example. Note the key features outlined below.

PROCESS SCHEMATICS – KEY FEATURES

- Simple line-drawn symbols are used to represent complex pieces of plant.
- The diagrams have a key showing each process element combination, linked to the plant area designation.
- Each schematic cross-references the other schematics that make up the complete 'set' for the process.

Useful references

http://www.topangasoftware.com/
http://infosys.kingston.ac.uk/ISSchool/Research/d.fischer/App-III.htm

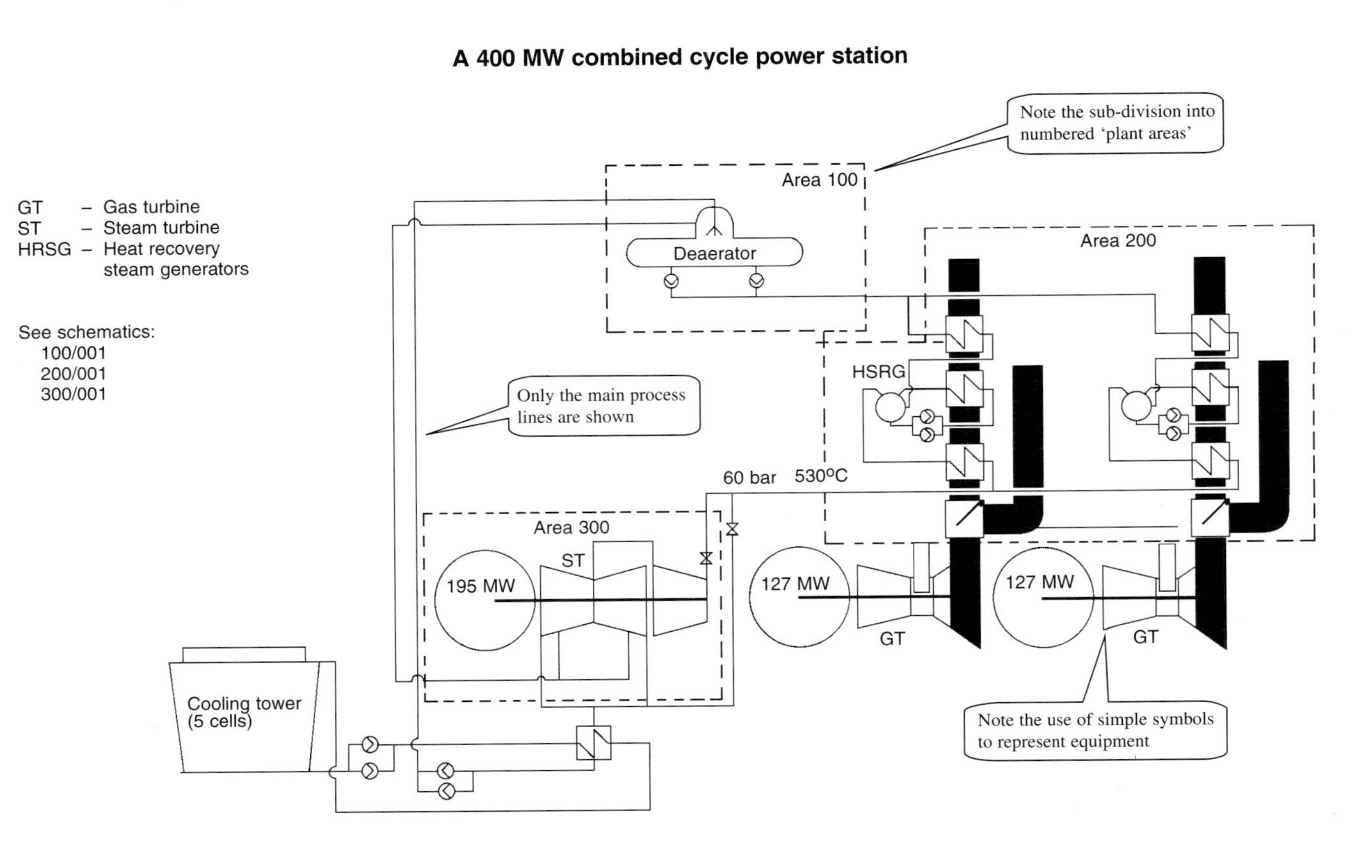

Fig. 12.1 A process schematic diagram

Process instrumentation diagrams (PIDs)

PIDs show the overall 'scheme' for the monitoring, instrumentation, and control of a flow process. To process engineers, this is key information about the character of the process and the implications for the mechanical and electrical engineering detail required to make it work.

PIDs are presented in the type of special format shown in Fig. 12.2 The purpose is to show the location of all the monitoring points, controlled variables, control links, and instruments – again, all in schematic form; because of the complexity of PIDs there will be many of them for each process schematic.

PIDs – KEY FEATURES

- Standard symbols are used for monitoring, control, and instrumentation components.
- Specific information on equipment ranges, etc. is given – engineers will use PIDs to specify the components to specialist sub-contractors.
- There is no reference to plant size or physical layout; PIDs are pure schematic diagrams.

Useful references

http://www.amgraf.com/02_d.html
http://www.catia.ibm.com/prodinfo/pid.html

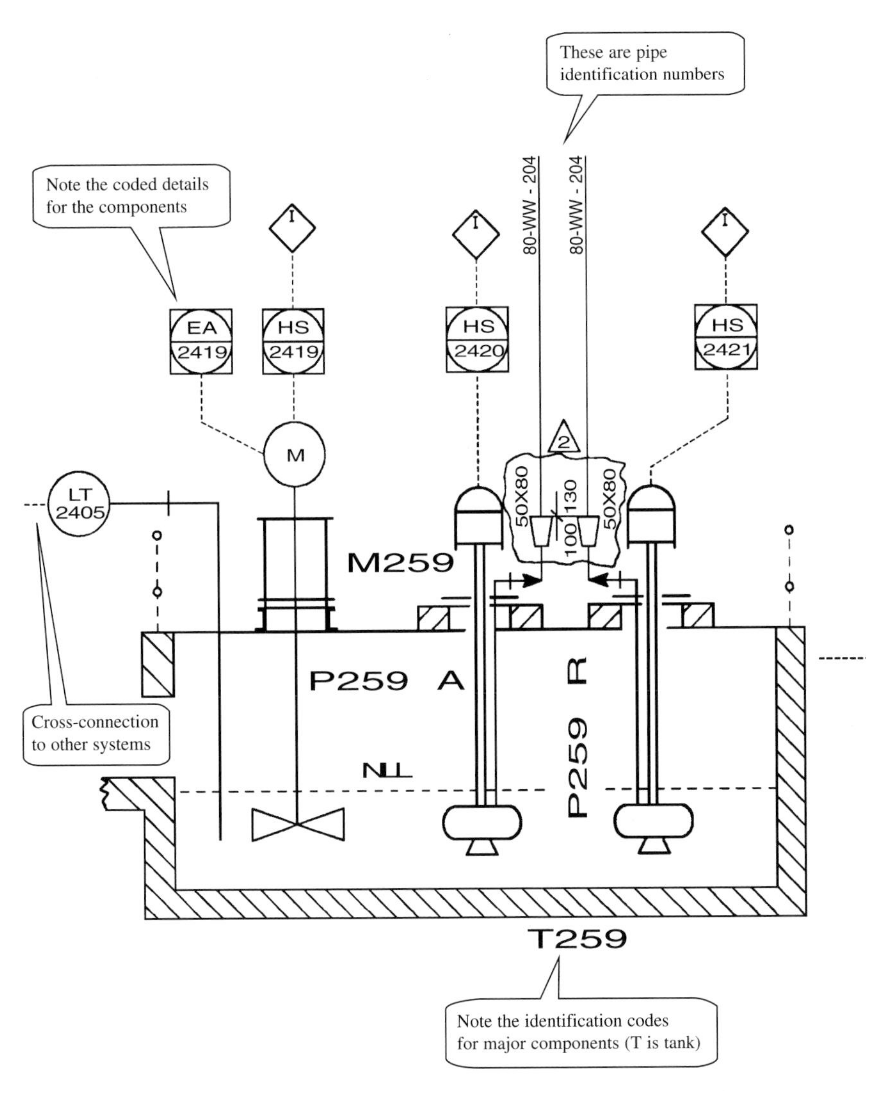

Fig. 12.2 A process instrumentation diagram (PID)

Flow/mass balance diagrams

These are an adaptation of the process schematic and show specific information on either the flow, mass, or thermal balance of a flow system. Plant components (or even complete areas) are represented by simple 'blocks' as shown in Fig. 12.3, with data on flow, energy, or other parameters shown on the links between the blocks.

FLOW/MASS BALANCE DIAGRAMS – KEY FEATURES

- The main purpose is to show balance and how flow or energy are apportioned within the process.
- It is essential to show a system boundary, as in Fig. 12.3. This can be a conceptual rather than physical boundary around part of a process, decided in order to help calculate thermal efficiencies or similar.
- It is normal to show the inputs to and outputs from the system outside the boundary.

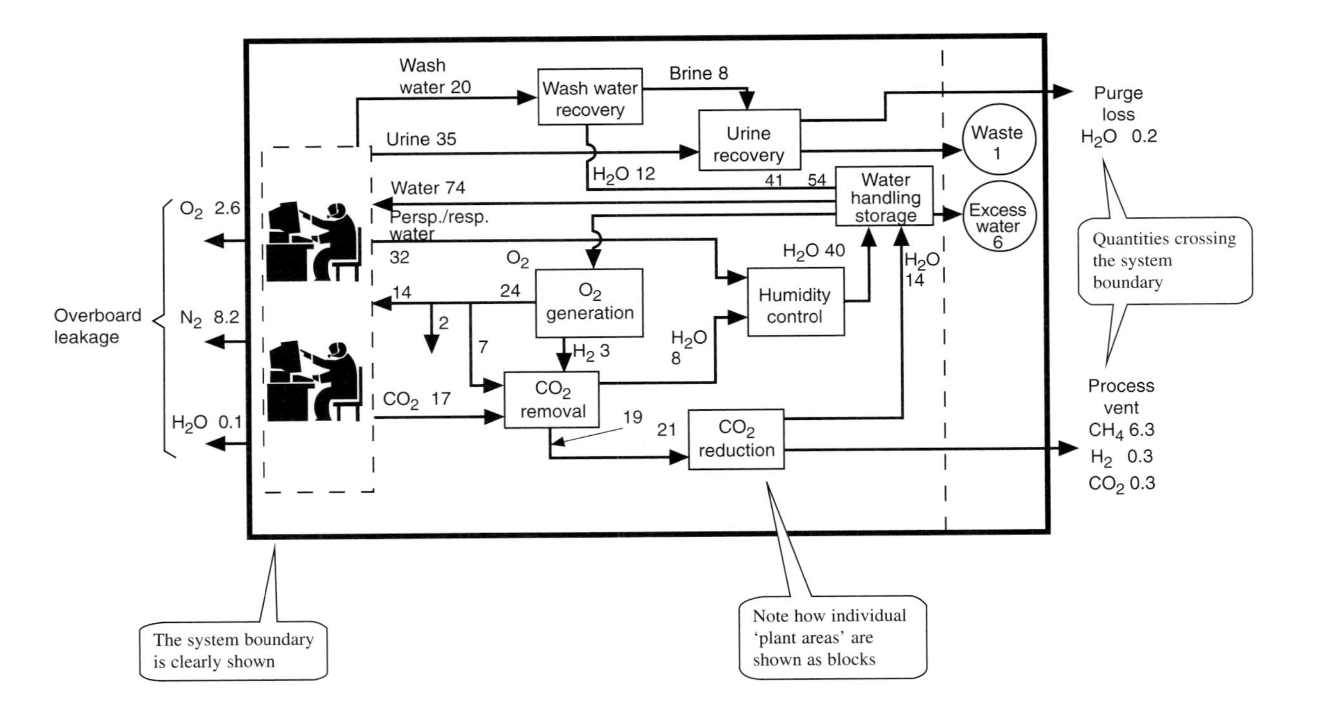

Fig. 12.3 A mass balance diagram

Piping diagrams

Piping diagrams follow on from the process schematics and PIDs. They come in several formats, but the most common is the type of hybrid schematic/isometric drawing shown in Fig. 12.4. The purpose of these drawings is to enable detailed specification and quantity (termed 'take off') information to be supplied to piping sub-contractors. The drawings normally show the pipe runs in isometric projection with the pipes drawn as single-thickness lines. Components such as valves, bends, reducers, and pipe supports are shown by symbols. Basic dimensions of the pipe runs are shown on the drawing which has an itemized key showing materials and sizes.

PIPING DIAGRAMS – KEY FEATURES

- Three-dimensional isometric presentation. Maximum of three or four pipe runs are shown per drawing.
- All pipes, valves, etc. are itemized and a basic description given in a table above or below the drawing.
- Conventions and symbols are listed in published technical standards.

Useful references

http://www.pipesite.com/powerbro1.html
http://software-guide.com/cdprod1/swhrec/003/309.shtml

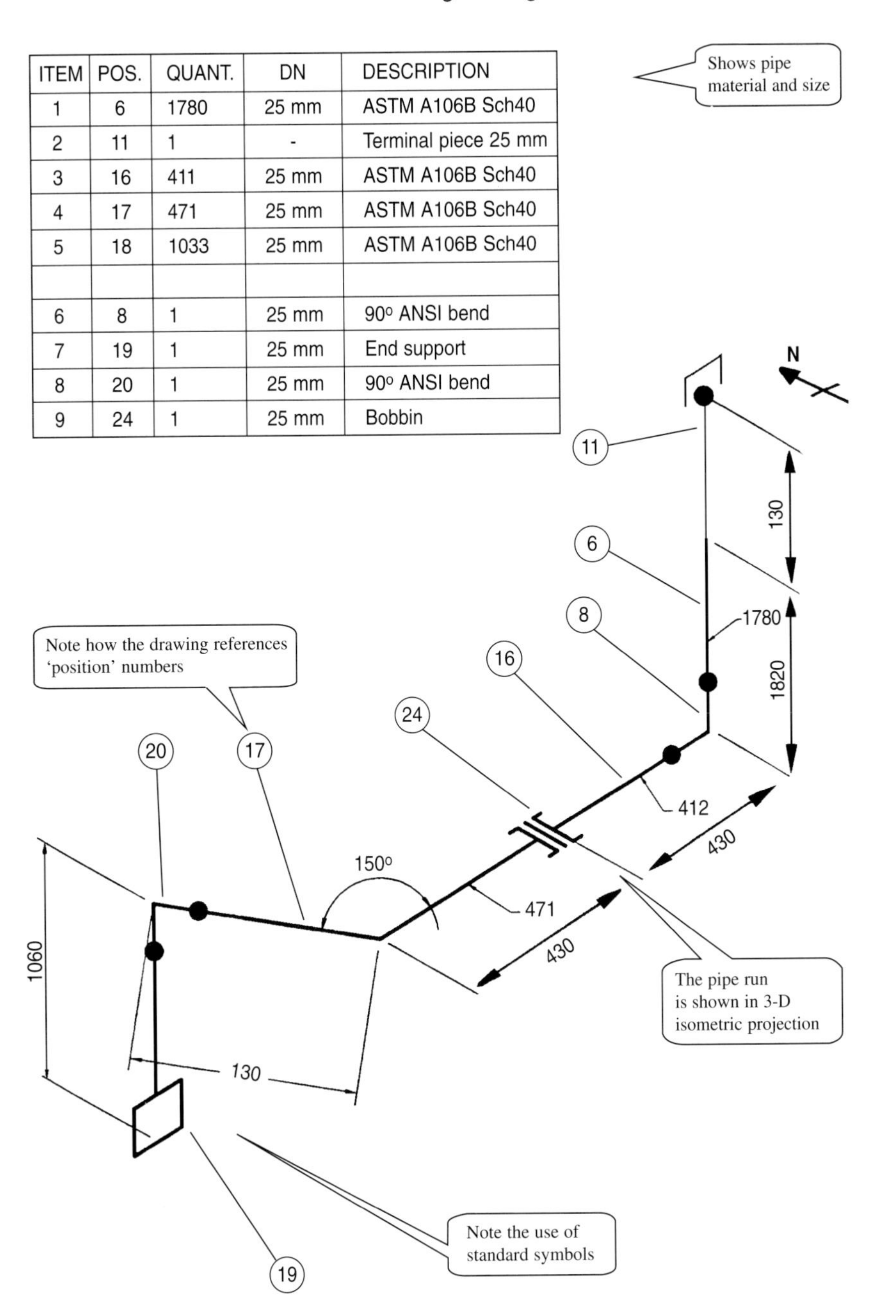

ITEM	POS.	QUANT.	DN	DESCRIPTION
1	6	1780	25 mm	ASTM A106B Sch40
2	11	1	-	Terminal piece 25 mm
3	16	411	25 mm	ASTM A106B Sch40
4	17	471	25 mm	ASTM A106B Sch40
5	18	1033	25 mm	ASTM A106B Sch40
6	8	1	25 mm	90° ANSI bend
7	19	1	25 mm	End support
8	20	1	25 mm	90° ANSI bend
9	24	1	25 mm	Bobbin

Fig. 12.4 A piping diagram

Pneumatic circuit diagrams

These are heavily stylized schematic diagrams that show the components of pneumatic circuits which are used in the control of a flow process or mechanical systems such as clutches and gearboxes. The diagram comprises large numbers of individual pneumatic and electro-pneumatic components such as solenoid valves, non-return valves, relays, etc., all of which are denoted by standard symbols. Figure 12.5 shows a typical example. Note the key features listed below.

PNEUMATIC CIRCUIT DIAGRAMS – KEY FEATURES

- Standard symbols are always used; be warned that there are hundreds of these, so you need to refer to the technical standards.
- Expect the diagrams to be complicated; they are not intended to be understood by non-technical people.
- Note how the drawings concentrate heavily on fail-safe features, reset conditions, and default settings. This is important information in pneumatic circuit design.

A pneumatic circuit for operation of a piston

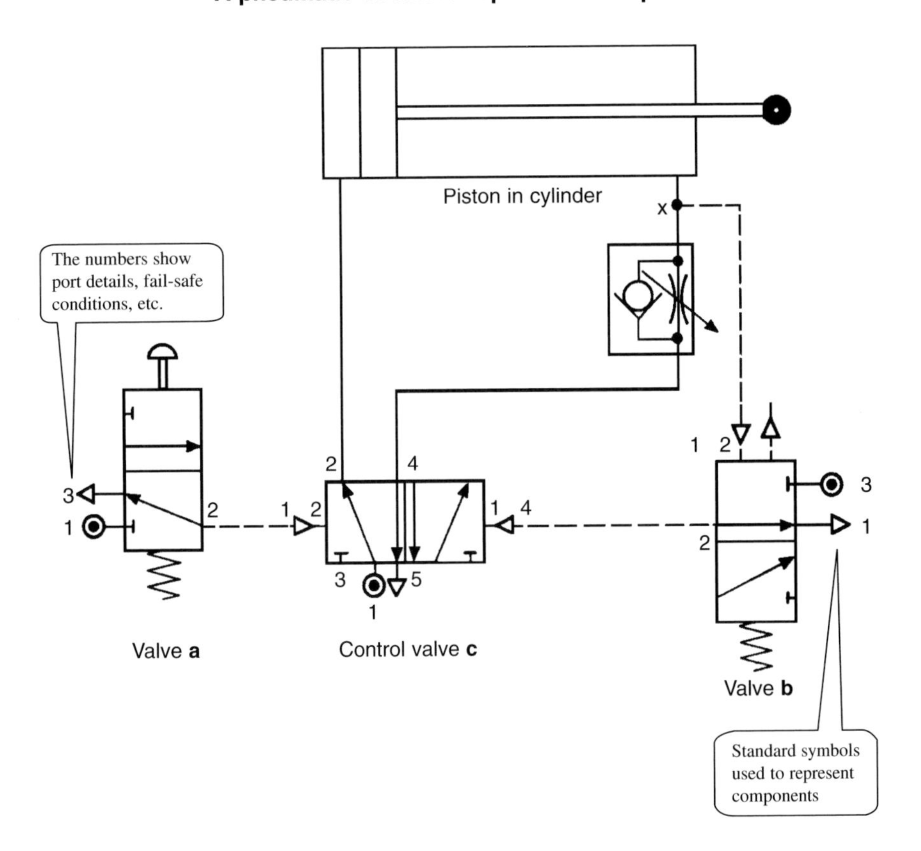

Fig. 12.5 A pneumatic circuit schematic

Terminal point diagrams (TPDs)

These are diagrams in which the sole purpose is to show points in a process system where one part of the construction contract ends and another begins. They are, therefore, primarily contract documents rather than technical ones. Figure 12.6 shows a typical example. Note the key features outlined below.

TERMINAL POINT DIAGRAMS – KEY FEATURES

- The purpose of TPDs is to show the allocation of contractual responsibilities for the various parts of a process plant.
- TPDs include all the parts of a plant and its processes, i.e. mechanical, electrical, control, utilities, etc.
- Some TPDs show guarantee parameters of process flows from one area of the plant to the next, so they form part of the commercial guarantee relating to the plant.

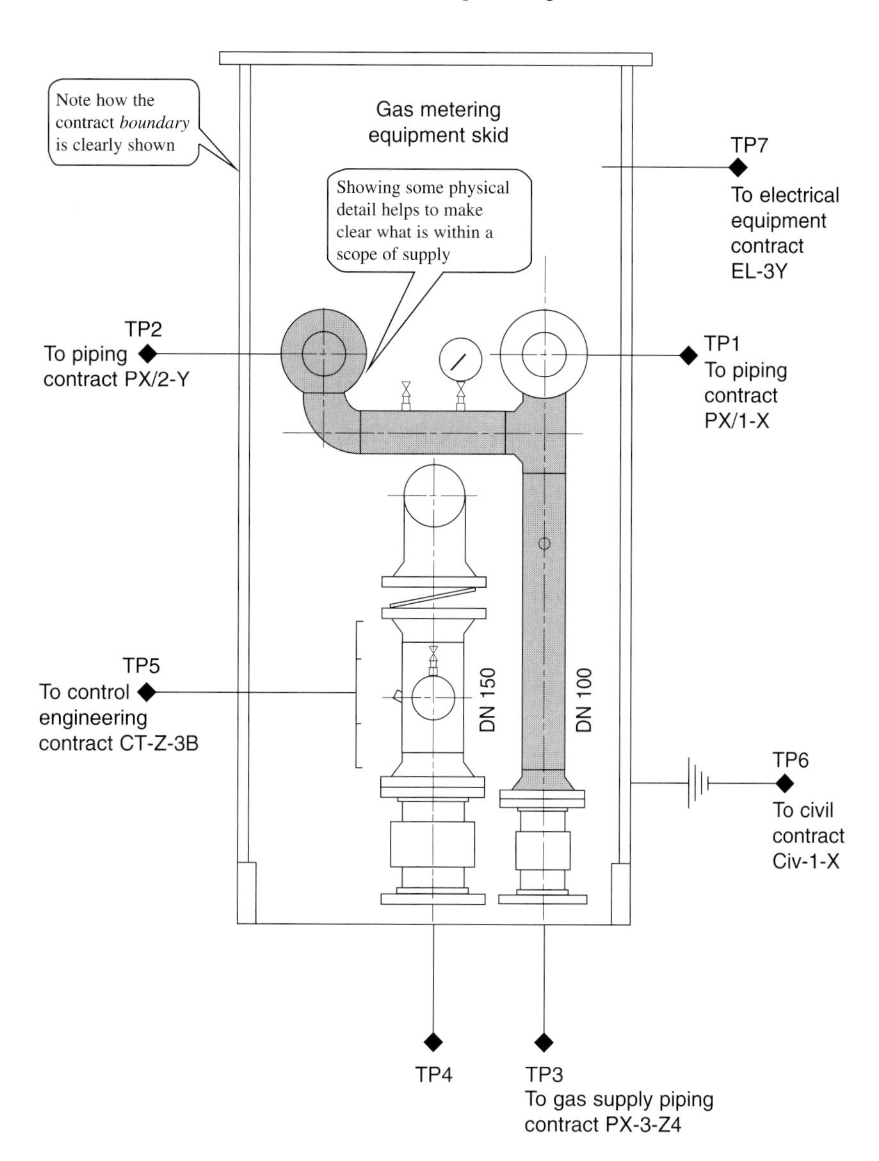

Fig. 12.6 A contract terminal point diagram

Site plans

Site plans, often called plot plans, provide an overall 'plan' view of a process plant or other industrial installation. They show the boundaries and dimensions for the main parts of the plant and its buildings and infrastructure. Landscape features, such as trees, are represented by symbols and the whole thing is drawn to a standard scale, normally 1:200 or 1:500, depending on the size of the site. Important information about underground services (water, drainage, electrical cables, etc.) is shown, even if it is not directly related to the plant under consideration. A larger scale version (1:1250 or 1:2500) is called a block plan and shows the plant site (shaded or hatched) in relation to the surrounding land and its features. Figure 12.7 shows typical examples.

(a) A site plan

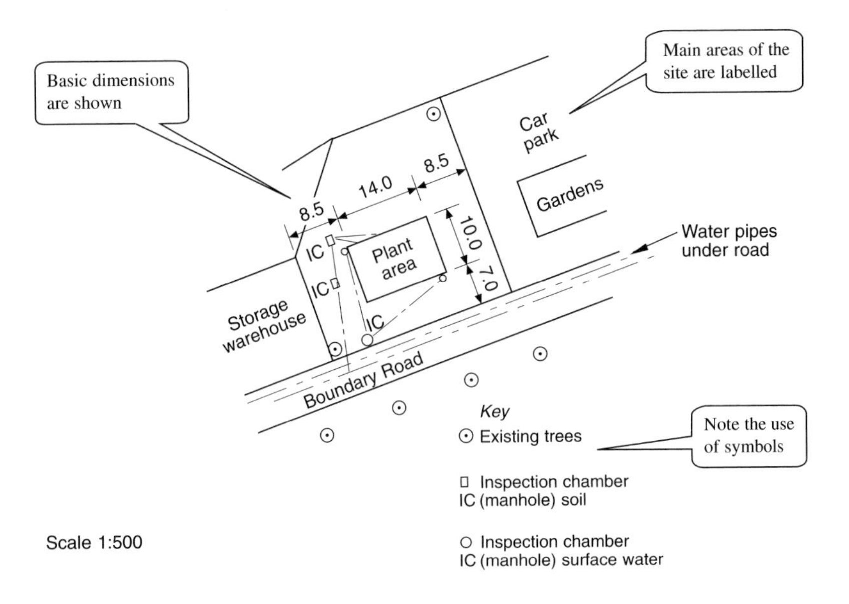

(b) A block plan

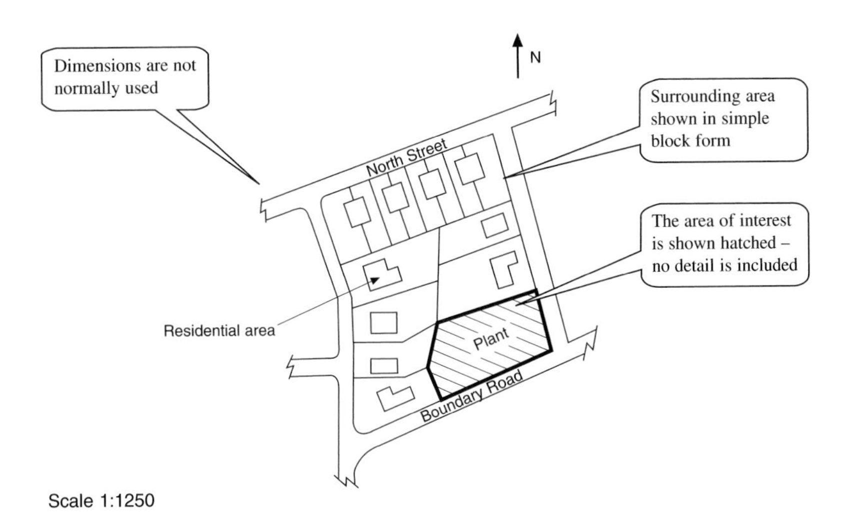

Fig. 12.7 Site and block plans

Floor plans

Floor plans are used for all technical disciplines that incorporate buildings or plant installations. The plan shows the view obtained by cutting horizontally through a building at about 1 metre above floor level, hence showing the position of external and internal windows, doors, etc. The scale is normally 1:50, 1:100 or 1:200 and the plans show all major dimensions. The floor plan is sometimes accompanied by a simplified vertical section drawing (actually more of a sketch) showing construction details of the foundations, walls, floor, and roof. It is convention to use various shading and hatching to represent materials of construction. Other types contain information in a more symbolic form intended 'for guidance only'. Figure 12.8 shows two typical examples.

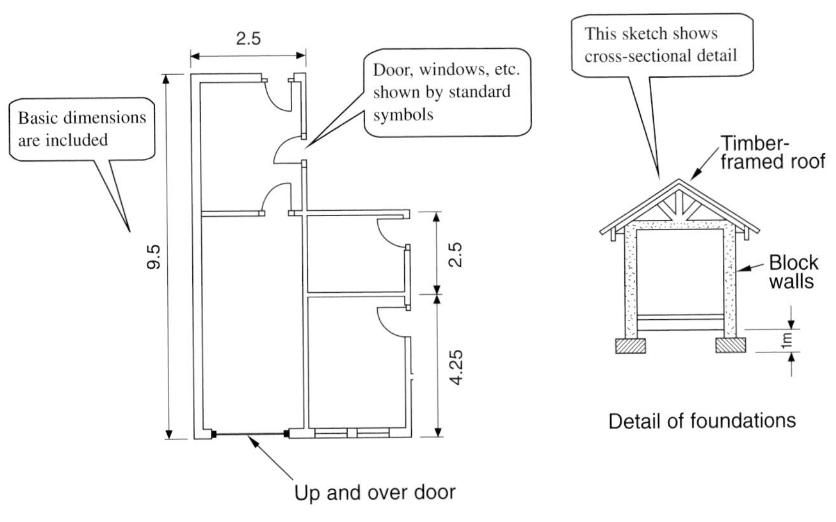

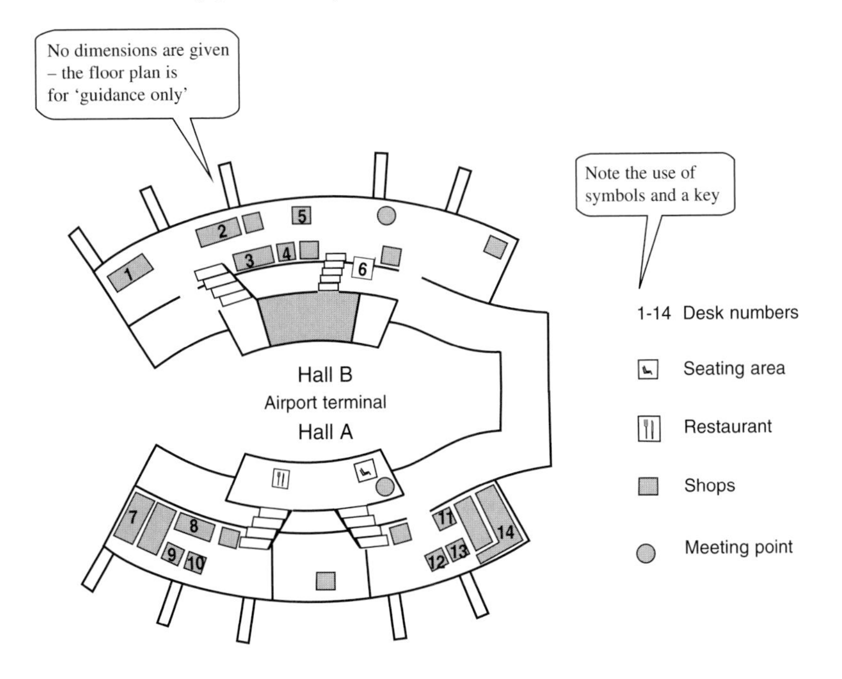

Fig. 12.8 Two types of floor plan drawing

Postscript – what now?

The fashion in presenting technical information is heading in the direction of more visual methods. In a world awash with technical information, people are still hungry for information presented in visual form, in a way which they can assimilate, even if they do not really understand the fine but esoteric mathematics behind it. In this book we have looked at many ways of showing technical things in a visual way, but what have we learned?

One central message is that technical diagrams are rarely restricted to use in a single technical discipline – they transfer well across disciplines without losing any of their effectiveness. We have also seen that many travel well across international boundaries, so can be understood by people of different languages and cultures without too much trouble. As a general rule of thumb, the higher the technical content of a presentation, the less room exists for misinterpretation because of the rigidity of the mathematical rules that support it.

Is there anything to be wary of? Only the dangers of convention. We saw in Chapters 3, 4, and 5 how many techniques of technical presentation are controlled closely by convention (engineering drawings are a good example). The danger with convention is always that it can stifle imagination, so you do need to guard against this. Imagination and innovation can make the dreariest of technical information interesting.

Finally, do not assume that all the possible methods of presenting technical information and diagrams are shown in this book. There are undoubtedly thousands more variations than are included here. The purpose of this book is to give you the basics; there is a lot more out there.

Clifford Matthews

Bibliography

French, M. (1992) *Form, Structure and Mechanism*, MacMillan Limited, UK.

Horton, W. K. (1991) *Illustrating Computer Documentation*, John Wiley and Sons, ISBN 047 1538450.

Pahl, G. and **Beirtz, W.** (1984) *Engineering Design*, Design Council Books, UK.

Price, J. (1993) *How to Communicate Technical Information*, Addison-Wesley, ISBN 0805 3682 99.

Tufte, E. R. (1992) *The Visual Display of Quantitative Information*, Graphics Press, ISBN 0961 392 10X.

Tufte, E. R. and **Krasng, D.** (1997) *Visual Explanations*, Graphics Press, ISBN 0961 392126.

Appendix

Technical standards mentioned in this book are available from the following source:

British Standards Institution (BSI)
British Standards House
389 Chiswick High Road
London
W4 4AL

Note that BSI is one of the few organizations in the UK permitted to purchase all international and most national standards direct from source, including ISO, IEC, DIN (English translations), ASME, and API.

Index